Agriculture for Beginners

by

Charles William Burkett and Frank Lincoln Stevens and Daniel Harvey Hill

PREFACE

Since its first publication "Agriculture for Beginners" has found a welcome in thousands of schools and homes. Naturally many suggestions as to changes, additions, and other improvements have reached its authors. Naturally, too, the authors have busied themselves in devising methods to add to the effectiveness of the book. Some additions have been made almost every year since the book was published. To embody all these changes and helpful suggestions into a strictly unified volume; to add some further topics and sections; to bring all farm practices up to the ideals of to-day; to include the most recent teaching of scientific investigators--these were the objects sought in the thorough revision which has just been given the book. The authors hope and think that the remaking of the book has added to its usefulness and attractiveness.

They believe now, as they believed before, that there is no line of separation between the science of agriculture and the practical art of agriculture. They are assured by the success of this book that agriculture is eminently a teachable subject. They see no difference between teaching the child the fundamental principles of farming and teaching the same child the fundamental truths of arithmetic, geography, or grammar. They hold that a youth should be trained for the farm just as carefully as he is trained for any other occupation, and that it is unreasonable to expect him to succeed without training.

If they are right in these views, the training must begin in the public schools. This is true for two reasons:

1. It is universally admitted that aptitudes are developed, tastes acquired, and life habits formed during the years that a child is in the public school. Hence, during these important years every child intended for the farm should be taught to know and love nature, should be led to form habits of observation, and should be required to begin a study of those great laws upon which agriculture is based. A training like this goes far toward making his life-work profitable and delightful.

2. Most boys and girls reared on a farm get no educational training except that given in the public schools. If, then, the truths that unlock the doors of nature are not taught in the public schools, nature and nature's laws will always be hid in night to a majority of our bread-winners. They must still in ignorance

and hopeless drudgery tear their bread from a reluctant soil.

The authors return hearty thanks to Professor Thomas F. Hunt, University of California; Professor Augustine D. Selby, Ohio Experiment Station; Professor W. F. Massey, horticulturist and agricultural writer; and Professor Franklin Sherman, Jr., State Entomologist of North Carolina, for aid in proofreading and in the preparation of some of the material.

CONTENTS

TO THE TEACHER

Teachers sometimes shrink from undertaking the teaching of a simple textbook on agriculture because they are not familiar with all the processes of farming. By the same reasoning they might hesitate to teach arithmetic because they do not know calculus or to teach a primary history of the United States because they are not versed in all history. The art of farming is based on the sciences dealing with the growth of plants and animals. This book presents in a simple way these fundamental scientific truths and suggests some practices drawn from them. Hence, even though many teachers may not have plowed or sowed or harvested, such teachers need not be embarrassed in mastering and heartily instructing a class in nature's primary laws.

If teachers realize how much the efficiency, comfort, and happiness of their pupils will be increased throughout their lives from being taught to cooperate with nature and to take advantage of her wonderful laws, they will eagerly begin this study. They will find also that their pupils will be actively interested in these studies bearing on their daily lives, and this interest will be carried over to other subjects. Whenever you can, take the pupils into the field, the garden, the orchard, and the dairy. Teach them to make experiments and to learn by the use of their own eyes and brains. They will, if properly led, astonish you by their efforts and growth.

You will find in the practical exercises many suggestions as to experiments that you can make with your class or with individual members. Do not neglect this first-hand teaching. It will be a delight to your pupils. In many cases it will be best to finish the experiments or observational work first, and later turn to the text to amplify the pupil's knowledge.

Although the book is arranged in logical order, the teacher ought to feel free to teach any topic in the season best suited to its study. Omit any chapter or section that does not bear on your crops or does not deal with conditions in your state.

The United States government and the different state experiment stations publish hundreds of bulletins on agricultural subjects. These are sent without cost, on application. It will be very helpful to get such of these bulletins as bear on the different sections of the book. These will be valuable additions to your school library. The authors would like to give a list of these bulletins bearing on each chapter, but it would soon be out of date, for the bulletins get out of print and are supplanted by newer ones. However, the United States Department of Agriculture prints a monthly list of its publications, and each state experiment station keeps a list of its bulletins. A note to the Secretary of Agriculture, Washington, D.C., or to your own state experiment station will promptly bring you these lists, and from them you can select what you need for your school.

AGRICULTURE FOR BEGINNERS

CHAPTER I

THE SOIL

SECTION I. ORIGIN OF THE SOIL

The word soil occurs many times in this little book. In agriculture this word is used to describe the thin layer of surface earth that, like some great blanket, is tucked around the wrinkled and age-beaten form of our globe. The harder and colder earth under this surface layer is called the subsoil. It should be noted, however, that in waterless and sun-dried regions there seems little difference between the soil and the subsoil.

Plants, insects, birds, beasts, men,--all alike are fed on what grows in this thin layer of soil. If some wild flood in sudden wrath could sweep into the ocean this earth-wrapping soil, food would soon become as scarce as it was in Samaria when mothers ate their sons. The face of the earth as we now see it, daintily robed in grass, or uplifting waving acres of corn, or even naked, water-scarred, and disfigured by man's neglect, is very different from what it was in its earliest days. How was it then? How was the soil formed?

Learned men think that at first the surface of the earth was solid rock. How was this rock changed into workable soil? Occasionally a curious boy picks up a rotten stone, squeezes it, and finds his hands filled with dirt, or soil. Now, just as the boy crumbled with his fingers this single stone, the great forces of nature with boundless patience crumbled, or, as it is called, disintegrated, the early rock mass. The simple but giant-strong agents that beat the rocks into powder with a clublike force a millionfold more powerful than the club force of Hercules were chiefly (1) heat and cold; (2) water, frost, and ice; (3) a very low form of vegetable life; and (4) tiny animals--if such minute bodies can be called animals. In some cases these forces acted singly; in others, all acted together to rend and crumble the unbroken stretch of rock. Let us glance at some of the methods used by these skilled soil-makers.

Heat and cold are working partners. You already know that most hot bodies shrink, or contract, on cooling. The early rocks were hot. As the outside shell of rock cooled from exposure to air and moisture it contracted. This shrinkage of the rigid rim of course broke many of the rocks, and here and there left cracks, or fissures. In these fissures water collected and froze. As freezing water expands with irresistible power, the expansion still further broke the rocks to pieces. The smaller pieces again, in the same way, were acted on by frost and ice and again crumbled. This process is still a means of soil-formation.

Running water was another giant soil-former. If you would understand its action, observe some usually sparkling stream just after a washing rain. The clear waters are discolored by mud washed in from the surrounding hills. As though disliking their muddy burden, the waters strive to throw it off. Here, as low banks offer chance, they run out into shallows and drop some of it. Here, as they pass a quiet pool, they deposit more. At last they reach the still water at the mouth of the stream, and there they leave behind the last of their mud load, and often form of it little three-sided islands called deltas. In the same way mighty rivers like the Amazon, the Mississippi, and the Hudson, when they are swollen by rain, bear great quantities of soil in their sweep to the seas. Some of the soil they scatter over the lowlands as they whirl seaward; the rest they deposit in deltas at their mouths. It is estimated that the Mississippi carries to the ocean each year enough soil to cover a square mile of surface to a depth of two hundred and sixty-eight feet.

The early brooks and rivers, instead of bearing mud, ran oceanward either bearing ground stone that they themselves had worn from the rocks by ceaseless fretting, or bearing stones that other forces had already dislodged. The large pieces were whirled from side to side and beaten against one another or against bedrock until they were ground into smaller and smaller pieces. The rivers distributed this rock soil just as the later rivers distribute muddy soil. For ages the moving waters ground against the rocks. Vast were the waters; vast the number of years; vast the results.

Glaciers were another soil-producing agent. Glaciers are streams "frozen and moving slowly but irresistibly onwards, down well-defined valleys, grinding and pulverizing the rock masses detached by the force and weight of their attack." Where and how were these glaciers formed?

Once a great part of upper North America was a vast sheet of ice. Whatever moisture fell from the sky fell as snow. No one knows what made this long winter of snow, but we do know that snows piled on snows until mountains of white were built up. The lower snow was by the pressure of that above it packed into ice masses. By and by some change of climate caused the masses of ice to break up somewhat and to move south and west. These moving masses, carrying rock and frozen earth, ground them to powder. King thus describes the stately movement of these snow mountains: "Beneath the bottom of this slowly moving sheet of ice, which with more or less difficulty kept itself conformable with the face of the land over which it was riding, the sharper outstanding points were cut away and the deeper river canyons filled in. Desolate and rugged rocky wastes were thrown down and spread over with rich soil."

The joint action of air, moisture, and frost was still another agent of soil-making. This action is called weathering. Whenever you have noticed the outside stones of a spring-house, you have noticed that tiny bits are crumbling from the face of the stones, and adding little by little to the soil. This is a slow way of making additions to the soil. It is estimated that it would take 728,000 years to wear away limestone rock to a depth of thirty-nine inches. But when you recall the countless years through which the weather has striven against the rocks, you can readily understand that its never-wearying activity has added immensely to the soil.

In the rock soil formed in these various ways, and indeed on the rocks themselves, tiny plants that live on food taken from the air began to grow. They grew just as you now see mosses and lichens grow on the surface of rocks. The decay of these plants added some fertility to the newly formed soil. The life and death of each succeeding generation of these lowly plants added to the soil matter accumulating on the rocks. Slowly but unceasingly the soil increased in depth until higher vegetable forms could flourish and add their dead bodies to it. This vegetable addition to the soil is generally known as humus.

In due course of time low forms of animal life came to live on these plants, and in turn by their work and their death to aid in making a soil fit for the plowman.

Thus with a deliberation that fills man with awe, the powerful forces of nature splintered the rocks, crumbled them, filled them with plant food, and turned their flinty grains into a soft, snug home for vegetable life.

SECTION II. TILLAGE OF THE SOIL

A good many years ago a man by the name of Jethro Tull lived in England. He was a farmer and a most successful man in every way. He first taught the English people and the world the value of thorough tillage of the soil. Before and during his time farmers did not till the soil very intelligently. They simply prepared the seed-bed in a careless manner, as a great many farmers do to-day, and when the crops were gathered the yields were not large.

Jethro Tull centered attention on the important fact that careful and thorough tillage increases the available plant food in the soil. He did not know why his crops were better when the ground was frequently and thoroughly tilled, but he knew that such tillage did increase his yield. He explained the fact by saying, "Tillage is manure." We have since learned the reason for the truth that Tull taught, and, while his explanation was incorrect, the practice that he was following was excellent. The stirring of the soil enables the air to circulate through it freely, and permits a breaking down of the compounds that contain the elements necessary to plant growth.

You have seen how the air helps to crumble the stone and brick in old buildings. It does the same with soil if permitted to circulate freely through it. The agent of the air that chiefly performs this work is called carbonic acid gas, and this gas is one of the greatest helpers the farmer has in carrying on his work. We must not forget that in soil preparation the air is just as important as any of the tools and implements used in cultivation.

If the soil is fertile and if deep plowing has always been done, good crops will result, other conditions being favorable. If, however, the tillage is poor, scanty harvests will always result. For most soils a two-horse plow is necessary to break up and pulverize the land.

A shallow soil can always be improved by properly deepening it. The principle of greatest importance in soil-preparation is the gradual deepening of the soil in order that plant-roots may have more comfortable homes. If the farmer has been accustomed to plow but four inches deep, he should adjust the plow so as to turn five inches at the next plowing, then six, and so on until the seed-bed is nine or ten inches deep. This gradual deepening will not injure the soil but will put it quickly in good condition. If to good tillage rotation of crops be added, the soil will become more fertile with each succeeding year.

The plow, harrow, and roller are all necessary to good tillage and to a proper preparation of the seed-bed. The soil must be made compact and clods of all sizes must be crushed. Then the air circulates freely, and paying crops are the rule and not the exception.

Tillage does these things: it increases the plant-food supply, destroys weeds, and influences the moisture content of the soil.

=EXERCISE=

1. What tools are used in tillage?

2. How should a poor and shallow soil be treated?

3. Why should a poor and shallow soil be well compacted before sowing the crop?

4. Explain the value of a circulation of air in the soil.

5. What causes iron to rust?

6. Why is a two-horse turning-plow better than a one-horse plow?

7. Where will clods do the least harm--on top of the soil or below the surface?

8. Do plant roots penetrate clods?

9. Are earthworms a benefit or an injury to the soil?

10. Name three things that tillage does.

SECTION III. THE MOISTURE OF THE SOIL

Did any one ever explain to you how important water is to the soil, or tell you why it is so important? Often, as you know, crops entirely fail because there is not enough water in the soil for the plants to drink. How necessary is it, then, that the soil be kept in the best possible condition to catch and hold enough water to carry the plant through dry, hot spells! Perhaps you are ready to ask, "How does the mouthless plant drink its stored-up water?"

The plant gets all its water through its roots. You have seen the tiny threadlike roots of a plant spreading all about in fine soil; they are down in the ground taking up plant food and water for the stalk and leaves above. The water, carrying plant food with it, rises in a simple but peculiar way through the roots and stems.

The plants use the food for building new tissue, that is, for growth. The water passes out through the leaves into the air. When the summers are dry and hot and there is but little water in the soil, the leaves shrink up. This is simply a method they have of keeping the water from passing too rapidly off into the air. I am sure you have seen the corn blades all shriveled on very hot days. This shrinkage is nature's way of diminishing the current of water that is steadily passing through the plant.

A thrifty farmer will try to keep his soil in such good condition that it will have a supply of water in it for growing crops when dry and hot weather comes. He can do this by deep plowing, by subsoiling, by adding any kind of decaying vegetable matter to the soil, and by growing crops that can be tilled frequently.

The soil is a great storehouse for moisture. After the clouds have emptied their waters into this storehouse, the water of the soil comes to the surface, where it is evaporated into the air. The water comes to the surface in just the same way that oil rises in a lamp-wick. This rising of the water is called capillarity.

It is necessary to understand what is meant by this big word. If into a pan of water you dip a glass tube, the water inside the tube rises above the level of the water in the pan. The smaller the tube the higher will the water rise. The greater rise inside is perhaps due to the fact that the glass attracts the particles of water more than the particles of water attract one another. Now apply this principle to the soil.

The soil particles have small spaces between them, and the spaces act just as the tube does. When the water at the surface is carried away by drying winds and warmth, the water deeper in the soil rises through the soil spaces. In this way water is brought from its soil storehouse as plants need it.

Of course when the underground water reaches the surface it evaporates. If we want to keep it for our crops, we must prepare a trap to hold it. Nature has shown us how this can be done. Pick up a plank as it lies on the ground. Under the plank the soil is wet, while the soil not covered by the plank is dry. Why? Capillarity brought the water to the surface, and the plank, by keeping away wind and warmth, acted as a trap to hold the moisture. Now of course a farmer cannot set a trap of planks over his fields, but he can make a trap of dry earth, and that will do just as well.

When a crop like corn or cotton or potatoes is cultivated, the fine, loose dirt stirred by the cultivating-plow will make a mulch that serves to keep water in the soil in the same way that the plank kept moisture under it. The mulch also helps to absorb the rains and prevents the water from running off the surface. Frequent cultivation, then, is one of the best possible ways of saving moisture. Hence the farmer who most frequently stirs his soil in the growing season, and especially in seasons of drought, reaps, other things being equal, a more abundant harvest than if tillage were neglected.

=EXERCISE=

1. Why is the soil wet under a board or under straw?

2. Will a soil that is fine and compact produce better crops than one that is loose and cloddy? Why?

3. Since the water which a plant uses comes through the roots, can the morning dew afford any assistance?

4. Why are weeds objectionable in a growing crop?

5. Why does the farmer cultivate growing corn and cotton?

SECTION IV. HOW THE WATER RISES IN THE SOIL

When the hot, dry days of summer come, the soil depends upon the subsoil, or undersoil, for the moisture that it must furnish its growing plants. The water was stored in the soil during the fall, winter, and spring months when there was plenty of rain. If you dig down into the soil when everything is dry and hot, you will soon reach a cool, moist undersoil. The moisture increases as you dig deeper into the soil.

Now the roots of plants go down into the soil for this moisture, because they need the water to carry the plant food up into the stems and leaves.

You can see how the water rises in the soil by performing a simple experiment.

=EXPERIMENT=

Take a lamp-chimney and fill it with fine, dry dirt. The dirt from a road or a field will do. Tie over the smaller end of the lamp-chimney a piece of cloth or a pocket handkerchief, and place this end in a shallow pan of water. If the soil in the lamp-chimney is clay and well packed, the water will quickly rise to the top.

By filling three or four lamp-chimneys with as many different soils, the pupil will see that the water rises more slowly in some than in others.

Now take the water pan away, and the water in the lamp-chimneys will gradually evaporate. Study for a few days the effect of evaporation on the several soils.

SECTION V. DRAINING THE SOIL

A wise man was once asked, "What is the most valuable improvement ever made in agriculture?" He answered, "Drainage." Often soils unfit for crop-production because they contain too much water are by drainage rendered the most valuable of farming lands.

Drainage benefits land in the following ways:

1. It deepens the subsoil by removing unnecessary water from the spaces between the soil particles. This admits air. Then the oxygen which is in the air, by aiding decay, prepares plant food for vegetation.

2. It makes the surface soil, or topsoil, deeper. It stands to reason that the deeper the soil the more plant food becomes available for plant use.

3. It improves the texture of the soil. Wet soil is sticky. Drainage makes this sticky soil crumble and fall apart.

4. It prevents washing.

5. It increases the porosity of soils and permits roots to go deeper into the soil for food and moisture.

6. It increases the warmth of the soil.

7. It permits earlier working in spring and after rains.

8. It favors the growth of germs which change the unavailable nitrogen of the soil into nitrates; that is, into the form of nitrogen most useful to plants.

9. It enables plants to resist drought better because the roots go into the ground deeper early in the season.

A soil that is hard and wet will not grow good crops. The nitrogen-gathering crops will store the greatest quantity of nitrogen in the soil when the soil is open to the free circulation of the air. These valuable crops cannot do this when the soil is wet and cold.

Sandy soils with sandy subsoils do not often need drainage; such soils are naturally drained. With clay soils it is different. It is very important to remove the stagnant water in them and to let the air in.

When land has been properly drained the other steps in improvement are easily taken. After soil has been dried and mellowed by proper drainage, then commercial fertilizers, barnyard manure, cowpeas, and clover can most readily do their great work of improving the texture of the soil and of making it fitter for plant growth.

=Tile Drains.= Tile drains are the best and cheapest that can be used. It would not be too strong to say that draining by tiles is the most perfect drainage. Thousands of practical tests in this country have proved the superiority of tile draining for the following reasons:

1. Good tile drains properly laid last for years and do not fill up.

2. They furnish the cheapest possible means of removing too much water from the soil.

3. They are out of reach of all cultivating tools.

4. Surface water in filtering through the tiles leaves its nutritious elements for plant growth.

=EXPERIMENTS=

=To show the Effect of Drainage.= Take two tomato cans and fill both with the same kind of soil. Punch several holes in the bottom of one to drain the soil above and to admit air circulation. Leave the other unpunctured. Plant seeds of any kind in both cans and keep in a warm place. Add every third day equal quantities of water. Let seeds grow in both cans and observe the difference in growth for two or three weeks.

=To show the Effect of Air in Soils.= Take two tomato cans; fill one with soil that is loose and warm, and the other with wet clay or muck from a swampy field. Plant a few seeds of the same kind in each and observe how much better the dry, warm, open soil is for growing farm crops.

SECTION VI. IMPROVING THE SOIL

We hear a great deal about the exhaustion or wearing out of the soil. Many uncomfortable people are always declaring that our lands will no longer produce profitable crops, and hence that farming will no longer pay.

Now it is true, unfortunately, that much land has been robbed of its fertility, and, because this is true, we should be most deeply interested in everything that leads to the improvement of our soils.

When our country was first discovered and trees were growing everywhere, we had virgin soils, or new soils that were rich and productive because they were filled with vegetable matter and plant food. There are not many virgin soils now because the trees have been cut from the best lands, and these lands have been farmed so carelessly that the vegetable matter and available plant food have been largely used up. Now that fresh land is scarce it is very necessary to restore fertility to these exhausted lands. What are some of the ways in which this can be done?

There are several things to be done in trying to reclaim worn-out land. One of the first of these is to till the land well. Many of you may have heard the story of the dying father who called his sons about him and whispered feebly, "There is great treasure hidden in the garden." The sons could hardly wait to bury their dead father before, thud, thud, thud, their picks were going in the garden. Day after day they dug; they dug deep; they dug wide. Not a foot of the crop-worn garden escaped the probing of the pick as the sons feverishly searched for the expected treasure. But no treasure was found. Their work seemed entirely useless.

"Let us not lose every whit of our labor; let us plant this pick-scarred garden," said the eldest. So the garden was planted. In the fall the hitherto neglected garden yielded a harvest so bountiful, so unexpected, that the meaning of their father's words dawned upon them. "Truly," they said, "a treasure was hidden there. Let us seek it in all our fields."

The story applies as well to-day as it did when it was first told. Thorough culture of the soil, frequent and intelligent tillage--these are the foundations of soil-restoration.

Along with good tillage must go crop-rotation and good drainage. A supply of organic matter will prevent heavy rains from washing the soil and carrying away plant food. Drainage will aid good tillage in allowing air to circulate between the soil particles and in arranging plant food so that plants can use it.

But we must add humus, or vegetable matter, to the soil. You remember that the virgin soils contained a great deal of vegetable matter and plant food, but by the continuous growing of crops like wheat, corn, and cotton, and by constant shallow tillage, both humus and plant food have been used up. Consequently much of our cultivated soil to-day is hard and dead.

There are three ways of adding humus and plant food to this lifeless land: the first way is to apply barnyard manure (to adopt this method means that livestock raising must be a part of all farming); the second way is to adopt rotation of crops, and frequently to plow under crops like clover and cowpeas; the third way is to apply commercial fertilizers.

To summarize: if we want to make our soil better year by year, we must cultivate well, drain well, and in the most economical way add humus and plant food.

=EXPERIMENT=

Select a small area of ground at your home and divide it into four sections, as shown in the following sketch:

On Section A apply barnyard manure; on Section B apply commercial fertilizers; on Section C apply nothing, but till well; on Section D apply nothing, and till very poorly.

A, B, and C should all be thoroughly plowed and harrowed. Then add barnyard manure to A, commercial fertilizers to B, and harrow A, B, and C at least four times until the soil is mellow and fine. D will most likely be cloddy, like many fields that we often see. Now plant on each plat some crop like cotton, corn, or wheat. When the plats are ready to harvest, measure the yield of each and determine whether the increased yield of the best plats has paid for the outlay for tillage and manure. The pupil will be much interested in the results obtained from the first crop.

Now follow a system of crop-rotation on the plats. Clover can follow corn or cotton or wheat; and cowpeas, wheat. Then determine the yield of each plat for the second crop. By following these plats for several years, and increasing the number, the pupils will learn many things of greatest value.

SECTION VII. MANURING THE SOIL

In the early days of our history, when the soil was new and rich, we were not compelled to use large amounts of manures and fertilizers. Yet our histories speak of an Indian named Squanto who came into one of the New England colonies and showed the first settlers how, by putting a fish in each hill of corn, they could obtain larger yields.

If people in those days, with new and fertile soils, could use manures profitably, how much more ought we to use them in our time, when soils have lost their virgin fertility, and when the plant food in the soil has been exhausted by years and years of cropping!

To sell year after year all the produce grown on land is a sure way to ruin it. If, for example, the richest land is planted every year in corn, and no stable or farmyard manure or other fertilizer returned to the soil, the land so treated will of course soon become too poor to grow any crop. If, on the other hand, clover or alfalfa or corn or cotton-seed meal is fed to stock, and the manure from the stock returned to the soil, the land will be kept rich. Hence those farmers who do not sell such raw products as cotton, corn, wheat, oats, and clover, but who market articles made from these raw products, find it easier to keep their land fertile. For illustration: if instead of selling hay, farmers feed it to sheep and sell meat and wool; if instead of selling cotton seed, they feed its meal to cows, and sell milk and butter; if instead of selling stover, they feed it to beef cattle, they get a good price for products and in addition have all the manure needed to keep their land productive and increase its value each year.

If we wish to keep up the fertility of our lands we should not allow anything to be lost from our farms. All the manures, straw, roots, stubble, healthy vines--in fact everything decomposable--should be plowed under or used as a top-dressing. Especial care should be taken in storing manure. It should be watchfully protected from sun and rain. If a farmer has no shed under which to keep his manure, he should scatter it on his fields as fast as it is made.

He should understand also that liquid manure is of more value than solid, because that important plant food, nitrogen, is found almost wholly in the liquid portion. Some of the phosphoric acid and considerable amounts of the potash are also found in the liquid manure. Hence economy requires that none of this escape either by leakage or by fermentation. Sometimes one can detect the smell of ammonia in the stable. This ammonia is formed by the decomposition of the liquid manure, and its loss should be checked by sprinkling some floats, acid phosphate, or muck over the stable floor.

Many farmers find it desirable to buy fertilizers to use with the manure made on the farm. In this case it is helpful to understand the composition, source, and availability of the various substances composing commercial fertilizers. The three most valuable things in commercial fertilizers are nitrogen, potash, and phosphoric acid.

The nitrogen is obtained from (1) nitrate of soda mined in Chile, (2) ammonium sulphate, a by-product of the gas works, (3) dried blood and other by-products of the slaughter-houses, and (4) cotton-seed meal. Nitrate of soda is soluble in water and may therefore be washed away before being used by plants. For this reason it should be applied in small quantities and at intervals of a few weeks.

Potash is obtained in Germany, where it is found in several forms. It is put on the market as muriate of potash, sulphate of potash, kainite, which contains salt as an impurity, and in other impure forms. Potash is found also in unleached wood ashes.

Phosphoric acid is found in various rocks of Tennessee, Florida, and South Carolina, and also to a large extent in bones. The rocks or bones are usually treated with sulphuric acid. This treatment changes the phosphoric acid into a form ready for plant use.

These three kinds of plant food are ordinarily all that we need to supply. In some cases, however, lime has to be added. Besides being a plant food itself, lime helps most soils by improving the structure of the grains; by sweetening the soil, thereby aiding the little living germs called bacteria; by hastening the decay of organic matter; and by setting free the potash that is locked up in the soil.

CHAPTER II

THE SOIL AND THE PLANT

SECTION VIII. ROOTS

You have perhaps observed the regularity of arrangement in the twigs and branches of trees. Now pull up the roots of a plant, as, for example, sheep sorrel, Jimson weed, or some other plant. Note the branching of the roots. In these there is no such regularity as is seen in the twig. Trace the rootlets to their finest tips. How small, slender, and delicate they are! Still we do not see the finest of them, for in taking the plant from the ground we tore the most delicate away. In order to see the real construction of a root we must grow one so that we may examine it uninjured. To do this, sprout some oats in a germinator or in any box in which one glass side has been arranged and allow the oats to grow till they are two or more inches high. Now examine the roots and you will see very fine hairs, similar to those shown in the accompanying figure, forming a fuzz over the surface of the roots near the tips. This fuzz is made of small hairs standing so close together that there are often as many as 38,200 on a single square inch. Fig. 17 shows how a root looks when it has been cut crosswise into what is known as a cross section. The figure is much increased in size. You can see how the root-hairs extend from the root in every direction. Fig. 18 shows a single root-hair very greatly enlarged, with particles of sand sticking to it.

These hairs are the feeding-organs of the roots, and they are formed only near the tips of the finest roots. You see that the large, coarse roots that you are familiar with have nothing to do with absorbing plant food from the soil. They serve merely to conduct the sap and nourishment from the root-hairs to the tree.

When you apply manure or other fertilizer to a tree, remember that it is far better to supply the fertilizer to the roots that are at some distance from the trunk, for such roots are the real feeders. The plant food in the manure soaks into the soil and immediately reaches the root-hairs. You can understand this better by studying the distribution of the roots of an orchard tree, shown in Fig. 19. There you can see that the fine tips are found at a long distance from the main trunk.

You can now readily see why it is that plants usually wilt when they are transplanted. The fine, delicate root-hairs are then broken off, and the plant can but poorly keep up its food and water supply until new hairs have been formed. While these are forming, water has been evaporating from the leaves, and consequently the plant does not get enough moisture and therefore droops.

Would you not conclude that it is very poor farming to till deeply any crop after the roots have extended between the rows far enough to be cut by the plow or cultivator? In cultivating between corn rows, for example, if you find that you are disturbing fine roots, you may be sure that you are breaking off millions of root-hairs from each plant and hence are doing harm rather than good. Fig. 20 shows how the roots from one corn row intertangle with those of another. You see at a glance how many of these roots would be destroyed by deep cultivation. Stirring the upper inch of soil when the plants are well grown is sufficient tillage and does no injury to the roots.

A deep soil is much better than a shallow soil, as its depth makes it just so much easier for the roots to seek deep food. Fig. 21 illustrates well how far down into the soil the alfalfa roots go.

=EXERCISE=

Dig up the roots of several cultivated plants and weeds and compare them. Do you find some that are fine or fibrous? some fleshy like the carrot? The dandelion is a good example of a tap-root. Tap-roots are deep feeders. Examine very carefully the roots of a medium-sized corn plant. Sift the dirt away gently so as to loosen as few roots as possible. How do the roots compare in area with the part above the ground? Try to trace a single root of the corn plant from the stalk to its very tip. How long are the roots of mature plants? Are they deep or shallow feeders? Germinate some oats or beans in a glass-sided box, as suggested, and observe the root-hairs.

SECTION IX. HOW THE PLANT FEEDS FROM THE SOIL

Plants receive their nourishment from two sources--from the air and from the soil. The soil food, or mineral food, dissolved in water, must reach the plant through the root-hairs with which all plants are provided in great numbers. Each of these hairs may be compared to a finger reaching among the particles of earth for food and water. If we examine the root-hairs ever so closely, we find no holes, or openings, in them. It is evident, then, that no solid particles can enter the root-hairs, but that all food must pass into the root in solution.

An experiment just here will help us to understand how a root feeds.

=EXPERIMENT=

Secure a narrow glass tube like the one in Fig. 22. If you cannot get a tube, a narrow, straight lamp-chimney will, with a little care, do nearly as well. From a bladder made soft by soaking, cut a piece large enough to cover the end of the tube or chimney and to hang over a little all around. Make the piece of bladder secure to the end of the tube by wrapping tightly with a waxed thread, as at B. Partly fill the tube with molasses (or it may be easier in case you use a narrow tube to fill it before attaching the bladder). Put the tube into a jar or bottle of water so placed that the level of the molasses inside and the water outside will be the same. Fasten the tube in this position and observe it frequently for three or four hours. At the end of the time you should find that the molasses in the tube has risen above the level of the liquid outside. It may even overflow at the top. If you use the lamp-chimney the rise will not be so clearly seen, since a greater volume is required to fill the space in the chimney. This increase in the contents of the tube is due to the entrance of water from the outside. The water has passed through the thin bladder, or membrane, and has come to occupy space in the tube. There is also a passage the other way, but the molasses can pass through the bladder membrane so slowly that the passage is scarcely noticeable. There are no holes, or openings, in the membrane, but still there is a free passage of liquids in both directions, although the more heavily laden solution must move more slowly.

A root-hair acts in much the same way as the tube in our experiment, with the exception that it is so made as to allow certain substances to pass in only one direction, that is, toward the inside. The outside of the root-hair is bathed in solutions rich in nourishment. The nourishment passes from the outside to the inside through the delicate membrane of the root-hair. Thus does food enter the plant-root. From the root-hairs, foods are carried to the inside of the root.

From this you can see how important it is for a plant to have fine, loose soil for its root-hairs; also how necessary is the water in the soil, since the food can be used only when it is dissolved in water.

This passage of liquids from one side of a membrane to another is called osmosis. It has many uses in the plant kingdom. We say a root takes nourishment by osmosis.

SECTION X. ROOT-TUBERCLES

Tubercle is a big word, but you ought to know how to pronounce it and what is meant by root-tubercles. We are going to tell you what a root-tubercle is and something about its importance to agriculture. When you have learned this, we are sure you will want to examine some plants for yourself in order that you may see just what tubercles look like on a real root.

Root-tubercles do not form on all kinds of plants that farmers grow. They are formed only on those kinds that botanists call legumes. The clovers, cowpeas, vetches, soy beans, and alfalfa are all legumes. The tubercles are little knotty, wart-like growths on the roots of the plants just named. These tubercles are caused by tiny forms of life called, as you perhaps already know, bacteria, or germs.

Instead of living in nests in trees like birds or in the ground like moles and worms, these tiny germs, less than one twenty-five thousandth of an inch long, make their homes on the roots of legumes. Nestling snugly together, they live, grow, and multiply in their sunless homes. Through their activity the soil is enriched by the addition of much nitrogen from the air. They are the good fairies of the farmer, and no magician's wand ever blessed a land so much as these invisible folk bless the land that they live in.

Just as bees gather honey from the flowers and carry it to the hives, where they prepare it for their own future use and for the use of others, so do these root-tubercles gather nitrogen from the air and fix it in their root homes, where it can be used by other crops.

In the earlier pages of this book you were told something about the food of plants. One of the main elements of plant food, perhaps you remember, is nitrogen. Just as soon as the roots of the leguminous plants begin to push down into the soil, the bacteria, or germs that make the tubercles, begin to build their homes on the roots, and in so doing they add nitrogen to the soil. You now see the importance of growing such crops as peas and clover on your land, for by their tubercles you can constantly add plant food to the soil. Now this much-needed nitrogen is the most costly part of the fertilizers that farmers buy every year. If every farmer, then, would grow these tubercle-bearing crops, he would rapidly add to the richness of his land and at the same time escape the necessity of buying so much expensive fertilizer.

=EXPERIMENT=

Take a spade or shovel and dig carefully around the roots of a cowpea and a clover plant; loosen the earth thoroughly and then pull the plants up, being careful not to break off any of the roots. Now wash the roots, and after they become dry count the nodules, or tubercles, on them. Observe the difference in size. How are they arranged? Do all leguminous plants have equal numbers of nodules? How do these nodules help the farmer?

SECTION XI. THE ROTATION OF CROPS

Doubtless you know what is meant by rotation, for your teacher has explained to you already how the earth rotates, or turns, on its axis and revolves around the sun. When we speak of crop-rotation we mean not only that the same crop should not be planted on the same land for two successive years but that crops should follow one another in a regular order.

Many farmers do not follow a system of farming that involves a change of crops. In some parts of the country the same fields are planted to corn or wheat or cotton year after year. This is not a good practice and sooner or later will wear out the soil completely, because the soil-elements that furnish the food of that constant crop are soon exhausted and good crop-production is no longer possible.

Why is crop-rotation so necessary? There are different kinds of plant food in the soil. If any one of these is used up, the soil of course loses its power to feed plants properly. Now each crop uses more of some of the different kinds of foods than others do, just as you like some kinds of food better than others. But the crop cannot, as you can, learn to use the kinds of food it does not like; it must use the kind that nature fitted it to use. Not only do different crops feed upon different soil foods, but they use different quantities of these foods.

Now if a farmer plant the same crop in the same field each year, that crop soon uses up all of the available plant food that it likes. Hence the soil can no longer properly nourish the crop that has been year by year robbing it. If that crop is to be successfully grown again on the land, the exhausted element must be restored.

This can be done in two ways: first, by finding out what element has here been exhausted, and then restoring this element by means either of commercial fertilizers or manure; second, by planting on the land crops that feed on different food and that will allow or assist kind Mother Nature "to repair her waste places." An illustration may help you to remember this fact. Nitrogen is, as already explained, one of the commonest plant foods. It may almost be called plant bread. The wheat crop uses up a good deal of nitrogen. Suppose a field were planted in wheat year after year. Most of the available nitrogen would be taken out of the soil after a while, and a new wheat crop, if planted on the field, would not get enough of its proper food to yield a paying harvest. This same land, however, that could not grow wheat could produce other crops that do not require so much nitrogen. For example, it could grow cowpeas. Cowpeas, aided by their root-tubercles, are able to gather from the air a great part of the nitrogen needed for their growth. Thus a good crop of peas can be obtained even if there is little available nitrogen in the soil. On the other hand wheat and corn and cotton cannot use the free nitrogen of the air, and they suffer if there is an insufficient quantity present in the soil; hence the necessity of growing legumes to supply what is lacking.

Let us now see how easily plant food may be saved by the rotation of crops.

If you sow wheat in the autumn it is ready to be harvested in time for planting cowpeas. Plow or disk the wheat stubble, and sow the same field to cowpeas. If the wheat crop has exhausted the greater part of the nitrogen of the soil, it makes no difference to the cowpea; for the cowpea will get its nitrogen from the air and not only provide for its own growth but will leave quantities of nitrogen in the queer nodules of its roots for the crops coming after it in the rotation.

If corn be planted, there should be a rotation in just the same way. The corn plant, a summer grower, of course uses a certain portion of the plant food stored in the soil. In order that the crop following the corn may feed on what the corn did not use, this crop should be one that requires a somewhat different food. Moreover, it should be one that fits in well with corn so as to make a winter crop. We find just such a plant in clover or wheat. Like the cowpea, all the varieties of clover have on their roots tubercles that add the important element, nitrogen, to the soil.

From these facts is it not clear that if you wish to improve your land quickly and keep it always fruitful you must practice crop-rotation?

AN ILLUSTRATION OF CROP-ROTATION

Here are two systems of crop-rotation as practiced at one or more agricultural experiment stations. Each furnishes an ideal plan for keeping up land.

FIRST YEAR		SECOND YEAR		THIRD YEAR	
Summer	Winter	Summer	Winter	Summer	Winter
Corn	Crimson clover	Cotton	Wheat	Cowpeas	Rye for pasture

or

Summer	Winter	Summer	Winter	Summer	Winter
Corn	Wheat and grass	Clover	Clover and grass	Grass for pasture or	Grass meadow

In these rotations the cowpeas and clovers are nitrogen-gathering crops. They not only furnish hay but they enrich the soil. The wheat, corn, and cotton are money crops, but in addition they are cultivated crops; hence they improve the physical condition of the soil and give opportunity to kill weeds. The grasses and clovers are of course used for pasturage and hay. This is only a suggested rotation. Work out one that will meet your home need.

Let the pupils each present a system of rotation that includes the crops raised at home. The system presented should as nearly as possible meet the following requirements:

1. Legumes for gathering nitrogen. 2. Money crops for cash income. 3. Cultivated crops for tillage and weed-destruction. 4. Food crops for feeding live stock.

CHAPTER III

THE PLANT

SECTION XII. HOW A PLANT FEEDS FROM THE AIR

If you partly burn a match you will see that it becomes black. This black substance into which the match changes is called carbon. Examine a fresh stick of charcoal, which is, as you no doubt know, burnt wood. You see in the charcoal every fiber that you saw in the wood itself. This means that every part of the plant contains carbon. How important, then, is this substance to the plant!

You will be surprised to know that the total amount of carbon in plants comes from the air. All the carbon that a plant gets is taken in by the leaves of the plant; not a particle is gathered by the roots. A large tree, weighing perhaps 11,000 pounds, requires in its growth carbon from 16,000,000 cubic yards of air.

Perhaps, after these statements, you may think there is danger that the carbon of the air may sometime become exhausted. The air of the whole world contains about 1,760,000,000,000 pounds of carbon. Moreover, this is continually being added to by our fires and by the breath of animals. When wood or coal is used for fuel the carbon of the burning substance is returned to the air in the form of gas. Some large factories burn great quantities of coal and thus turn much carbon back to the air. A single factory in Germany is estimated to give back to the air daily about 5,280,000 pounds of carbon. You see, then, that carbon is constantly being put back into the air to replace that which is used by growing plants.

The carbon of the air can be used by none but green plants, and by them only in the sunlight. We may compare the green coloring matter of the leaf to a machine, and the sunlight to the power, or energy, which keeps the machine in motion. By means, then, of sunlight and the green coloring matter of the leaves, the plant secures carbon. The carbon passes into the plant and is there made into two foods very necessary to the plant; namely, starch and sugar.

Sometimes the plant uses the starch and sugar immediately. At other times it stores both away, as it does in the Irish and the sweet potato and in beets, cabbage, peas, and beans. These plants are used as food by man because they contain so much nourishment; that is, starch and sugar which were stored away by the plant for its own future use.

=EXERCISE=

Examine some charcoal. Can you see the rings of growth? Slightly char paper, cloth, meat, sugar, starch, etc. What does the turning black prove? What per cent of these substances do you think is pure carbon?

SECTION XIII. THE SAP CURRENT

The root-hairs take nourishment from the soil. The leaves manufacture starch and sugar. These manufactured foods must be carried to all parts of the plant. There are two currents to carry them. One passes from the roots through the young wood to the leaves, and one, a downward current, passes through the bark, carrying needed food to the roots (see Fig. 28).

If you should injure the roots, the water supply to the leaves would be cut off and the leaves would immediately wither. On the other hand, if you remove the bark, that is, girdle the tree, you in no way interfere with the water supply and the leaves do not wither. Girdling does, however, interfere with the downward food current through the bark.

If the tree be girdled the roots sooner or later suffer from lack of food supply from the leaves. Owing to this food stoppage the roots will cease to grow and will soon be unable to take in sufficient water, and then the leaves will begin to droop. This, however, may not happen until several months after the girdling. Sometimes a partly girdled branch grows much in thickness just above the girdle, as is shown in Fig. 29. This extra growth seems to be due to a stoppage of the rich supply of food which was on its way to the roots through the bark. It could go no farther and was therefore used by the tree to make an unnatural growth at this point. You will now understand how and why trees die when they are girdled to clear new ground.

It is, then, the general law of sap-movement that the upward current from the roots passes through the woody portion of the trunk, and that the current bearing the food made by the leaves passes downward through the bark.

=EXERCISE=

Let the teacher see that these and all other experiments are performed by the pupils. Do not allow them to guess, but make them see.

Girdle valueless trees or saplings of several kinds, cutting the bark away in a complete circle around the tree. Do not cut into the wood. How long before the tree shows signs of injury? Girdle a single small limb on a tree. What happens? Explain.

SECTION XIV. THE FLOWER AND THE SEED

Some people think that the flowers by the wayside are for the purpose of beautifying the world and increasing man's enjoyment. Do you think this is true? Undoubtedly a flower is beautiful, and to be beautiful is one of the uses of many flowers; but it is not the chief use of a flower.

You know that when peach or apple blossoms are nipped by the spring frost the fruit crop is in danger. The fruit of the plant bears the seed, and the flower produces the fruit. That is its chief duty.

Do you know any plant that produces seed without flowers? Some one answers, "The corn, the elm, and the maple all produce seed, but have no flower." No, that is not correct. If you look closely you will find in the spring very small flowers on the elm and on the maple, while the ear and the tassel are really the blossoms of the corn plant. Every plant that produces seed has flowers, although they may sometimes seem very curious flowers.

Let us see what a flower really is. Take, for example, a buttercup, cotton, tobacco, or plum blossom (see Figs. 31 and 32). You will find on the outside a row of green leaves inclosing the flower when it is still a bud. These leaves are the sepals. Next on the inside is a row of colored leaves, or petals. Arranged inside of the petals are some threadlike parts, each with a knob on the end. These are the stamens. Examine one stamen closely (Fig. 33). On the knob at its tip you should find, if the flower is fully open, some fine grains, or powder. In the lily this powder is so abundant that in smelling the flower you often brush a quantity of it off on your nose. This substance is called pollen, and the knob on the end of the stamen, on which the pollen is borne, is the anther.

The pollen is of very great importance to the flower. Without it there could be no seeds. The stamens as pollen-bearers, then, are very important. But there is another part to each flower that is of equal value. This part you will find in the center of the flower, inside the circle of stamens. It is called the pistil (Fig. 32). The swollen tip of the pistil is the stigma. The swollen base of the pistil forms the ovary. If you carefully cut open this ovary you will find in it very small immature seeds.

Some plants bear all these parts in the same flower; that is, each blossom has stamens, pistil, petals, and sepals. The pear blossom and the tomato blossom represent such flowers. Other plants bear their stamens and pistils in separate blossoms. Stamens and pistils may even occur in separate plants, and some blossoms have no sepals or petals at all. Look at the corn plant. Here the tassel is a cluster of many flowers, each of which bears only stamens. The ear is likewise a cluster of many flowers, each of which bears only a pistil. The dust that you see falling from the tassel is the pollen, and the long silky threads of the ear are the stigmas.

Now no plant can bear seeds unless the pollen of the stamen falls on the stigma. Corn cannot therefore form seed unless the dust of the tassel falls upon the silk. Did you ever notice how poorly the cob is filled on a single cornstalk standing alone in a field? Do you see why? It is because when a plant stands alone the wind blows the pollen away from the tassel, and little or none is received on the stigmas below.

In the corn plant the stamens and pistils are separate; that is, they do not occur on the same flower, although they are on the same plant. This is also true of the cucumber (see Fig. 35). In many plants, however, such as the hemp, hop, sassafras, willow, and others, the staminate parts are on one plant and the pistillate parts are on another. This is also true in several other cultivated plants. For example, in some strawberries the stamens are absent or useless; that is, they bear no good pollen. In such cases the grower must see to it that near by are strawberry plants that bear stamens, in order that those plants which do not bear pollen may become pollinated; that is, may have pollen carried to them. After the stigma has been supplied with pollen, a single pollen grain sends a threadlike sprout down through the stigma into the ovary. This process, if successfully completed, is called fertilization.

=EXERCISE=

Examine several flowers and identify the parts named in the last section. Try in the proper season to find the pollen on the maple, willow, alder, and pine, and on wheat, cotton, and the morning-glory.

How fast does the ovary of the apple blossom enlarge? Measure one and watch it closely from day to day. Can you find any plants that have their stamens and ovaries on separate individuals?

SECTION XV. POLLINATION

Nature has several interesting ways of bringing about pollination. In the corn, willow, and pine the pollen is picked up by the wind and carried away. Much of it is lost, but some reaches the stigmas, or receptive parts, of other corn, willow, or pine flowers. This is a very wasteful method, and all plants using it must provide much pollen.

Many plants employ a much better method. They have learned how to make insects bear their pollen. In plants of this type the parts of the blossom are so shaped and so placed as to deposit pollen from the stamen on the insect and to receive pollen from the insect on the stigmas.

When you see the clumsy bumblebee clambering over and pushing his way into a clover blossom, you may be sure that he is getting well dusted with pollen and that the next blossom which he visits will secure a full share on its stigmas.

When flowers fit themselves to be pollinated by insects they can no longer use the wind and are helpless if insects do not visit them. They therefore cunningly plan two ways to invite the visits of insects. First, they provide a sweet nectar as a repast for the insect visitor. The nectar is a sugary solution found in the bottom of the flower and is used by the visitor as food or to make honey. Second, flowers advertise to let each insect know that they have something for it. The advertising is done either by showy colors or by perfume. Insects have wonderful powers of smell. When you see showy flowers or smell fragrant ones, you will know that such flowers are advertising the presence either of nectar or of pollen (to make beebread) and that such flowers depend on insects for pollination.

A season of heavy, cold rains during blossoming-time may often injure the fruit crop by preventing insects from carrying pollen from flower to flower. You now also understand why plants often fail to produce seeds indoors. Since they are shut in, they cannot receive proper insect visits. Plants such as tomatoes or other garden fruits dependent upon insect pollination must, if raised in the greenhouse where insects cannot visit them, be pollinated by hand.

=EXERCISE=

Exclude insect visitors from some flower or flower cluster, for example, clover, by covering with a paper bag, and see whether the flower can produce seeds that are capable of growing. Compare as to number and vitality the seeds of such a flower with those of an uncovered flower. Observe insects closely. Do you ever find pollen on them? What kinds of insects visit the clover? the cowpea? the sourwood? the flax? Is wheat pollinated by insects or by the wind or by some other means? Do bees fly in rainy weather? How will a long rainy season at blossoming-time affect the apple crop? Why? Should bees be kept in an orchard? Why?

SECTION XVI. CROSSES, HYBRIDS, AND CROSS-POLLINATION

In our study of flowers and their pollination we have seen that the seed is usually the descendant of two parents, or at least of two organs--one the ovary, producing the seed; the other the pollen, which is necessary to fertilize the ovary.

It happens that sometimes the pollen of one blossom fertilizes the ovary of its own flower, but more often the pollen from one plant fertilizes the ovary of another plant. This latter method is called cross-pollination. As a rule cross-pollination makes seed that will produce a better plant than simple pollination would. Cross-pollination by hand is often used by plant-breeders when, for purposes of seed-selection, a specially strong plant is desired. The steps in hand pollination are as follows: (1) remove the anthers before they open, to prevent them from pollinating the stigma (the steps in this process are illustrated in Figs. 37, 38-39); (2) cover the flower thus treated with a paper bag to prevent stray pollen from getting on it (see Fig. 40); (3) when the ovary is sufficiently developed, carry pollen to the stigma by hand from the anthers of another plant which you have selected to furnish it, and rebag to keep out any stray pollen which might accidentally get in; (4) collect the seeds when they are mature and label them properly.

Hand pollination has this advantage--you know both parents of your seed. If pollination occur naturally you know the maternal but have no means of judging the paternal parent. You can readily see, therefore, how hand pollination enables you to secure seed derived from two well-behaved parents.

Sometimes we can breed one kind of plant on another. The result of such cross-breeding is known as a hybrid. In the animal kingdom the mule is a common example of this cross-breeding. Plant hybrids were formerly called mules also, but this suggestive term is almost out of use.

It is only when plants of two distinct kinds are crossed that the result is called a hybrid; for example, a blackjack oak on a white oak, an apple on a pear. If the parent plants are closely related, for example, two kinds of apples, the resulting plant is known simply as a cross.

Hybrids and crosses are valuable in that they usually differ from both parents and yet combine some qualities of each.

They often leave off some of the qualities of the parent plants and at other times have such qualities more markedly than did their parents. Thus they often produce an interesting new kind of plant. Sometimes we are able by hybridization to combine in one plant the good qualities of two other plants and thus make a great advance in agriculture. The new forms brought about by hybridization may be fixed, or made permanent, by such selection as is mentioned in Section XVIII. Hybridization is of great aid in originating new plants.

It often happens that a plant will be more fruitful when pollinated by one variety than by some other variety. This is well illustrated in Fig. 41. A fruit-grower or farmer should know much about these subjects before selecting varieties for his orchard, vineyard, etc.

=EXERCISE=

With the help of your teacher try to cross some plants. Such an experiment will take time, but will be most interesting. You must remember that many crosses must be attempted in order to gain success with even a few.

SECTION XVII. PROPAGATION BY BUDS

It is the business of the farmer to make plants grow, or, as it is generally called, to propagate plants. This he does in one of two ways: by buds (that is, by small pieces cut from parent plants), or by seeds. The chief aim in both methods should be to secure in the most convenient manner the best-paying plants.

Many plants are most easily and quickly propagated by buds; for example, the grape, red raspberry, fig, and many others that we cultivate for the flower only, such as the carnation, geranium, rose, and begonia.

In growing plants from cuttings, a piece is taken from the kind of plant that one wishes to grow. The greatest care must be exercised in order to get a healthy cutting. If we take a cutting from a poor plant, what can we expect but to grow a poor plant like the one from which our cutting was taken? On the other hand, if a fine, strong, vigorous, fruitful plant be selected, we shall expect to grow just such a fine, healthy, fruitful plant.

We expect the cutting to make exactly the same variety of plant as the parent stock. We must therefore decide on the variety of berry, grape, fig, carnation, or rose that we wish to propagate, and then look for the strongest and most promising plants of this variety within our reach. The utmost care will not produce a fine plant if we start from poor stock.

What qualities are most desirable in a plant from which cuttings are to be taken? First, it should be productive, hardy, and suited to your climate and your needs; second, it should be healthy. Do not take cuttings from a diseased plant, since the cutting may carry the disease.

Cuttings may be taken from various parts of the plant, sometimes even from parts of the leaf, as in the begonia (Fig. 46). More often, however, they are drawn from parts of the stem (Figs. 43-45). As to the age of the twig from which the cutting is to be taken, Professor Bailey says: "For most plants the proper age or maturity of wood for the making of cuttings may be determined by giving the twig a quick bend; if it snaps and hangs by the bark, it is in proper condition. If it bends without breaking, it is too young and soft or too old. If it splinters, it is too old and woody." Some plants, as the geranium (Fig. 42), succeed best if the cuttings from which they are grown are taken from soft, young parts of the plant; others, for example, the grape or rose, do better when the cutting is made from more mature wood.

Cuttings may vary in size and may include one or more buds. After a hardy, vigorous cutting is made, insert it about one half or one third of its length in soil. A soil free from organic matter is much the best, since in such soil the cuttings are much less liable to disease. A fine, clean sand is commonly used by professional gardeners. When cuttings have rooted well--this may require a month or more--they may be transplanted to larger pots.

Sometimes, instead of cutting off a piece and rooting it, portions of branches are made to root before they are separated from the parent plant. This method is often followed, and is known as layering. It is a simple process. Just bend the tip of a bough down and bury it in the earth (see Fig. 47). The black raspberry forms layers naturally, but gardeners often aid it by burying the over-hanging tips in the earth, so that more tips may easily take root. Strawberries develop runners that root themselves in a similar fashion.

Grafts and buds are really cuttings which, instead of being buried in sand to produce roots of their own, are set on the roots of other plants.

Grafting and budding are practiced when these methods are more convenient than cuttings or when the gardener thinks there is danger of failure to get plants to take root as cuttings. Neither grafting nor budding is, however, necessary for the raspberry or the grape, for these propagate most readily from cuttings.

It is often the case that a budded or grafted plant is more fruitful than a plant on its own roots. In cases of this kind, of course, grafts or buds are used.

The white, or Irish, potato is usually propagated from pieces of the potato itself. Each piece used for planting bears one eye or more. The potato itself is really an underground stem and the eyes are buds. This method of propagation is therefore really a peculiar kind of cutting.

Since the eye is a bud and our potato plant for next year is to develop from this bud, it is of much importance, as we have seen, to know exactly what kind of plant our potato comes from. If the potato is taken from a small plant that had but a few poor potatoes in the hill, we may expect the bud to produce a similar plant and a correspondingly poor crop. We must see to it, then, that our seed potatoes are drawn from vines that were good producers, because new potato plants are like the plants from which they were grown. Of course when our potatoes are in the bin we cannot tell from what kind of plants they came. We must therefore select our seed potatoes in the field. Seed potatoes should always be selected from those hills that produce most bountifully. Be assured that the increased yield will richly repay this care in selecting. It matters not so much whether the seed potato be large or small; it must, however, come from a hill bearing a large yield of fine potatoes.

Sweet-potato plants are produced from shoots, or growing buds, taken from the potato itself, so that in their case too the piece that we use in propagating is a part of the original plant, and will therefore be like it under similar conditions. Just as with the Irish potato, it is important to know how good a yielder you are planting. You should watch during harvest and select for propagation for the next year only such plants as yield best.

We should exercise fully as much care in selecting proper individuals from which to make a cutting or a layer as we do in selecting a proper animal to breed from. Just as we select the finest Jersey in the herd for breeding purposes, so we should choose first the variety of plant we desire and then the finest individual plant of that variety.

If the variety of the potato that we desire to raise be Early Rose, it is not enough to select any Early Rose plants, but the very best Early Rose plants, to furnish our seed.

It is not enough to select large, fine potatoes for cuttings. A large potato may not produce a bountifully yielding plant. It will produce a plant like the one that produced it. It may be that this one large potato was the only one produced by the original plant. If so, the plant that grows from it will tend to be similarly unproductive. Thus you see the importance of selecting in the field a plant that has exactly the qualities desired in the new plant.

One of the main reasons why gardeners raise plants from buds instead of from seeds is that the seed of many plants will not produce plants like the parent. This failure to "come true," as it is called, is sometimes of value, for it occasionally leads to improvement. For example, suppose that a thousand apple or other fruit or flower seeds from plants usually propagated by cuttings be planted; it may be that one out of a thousand or a million will be a very valuable plant. If a valuable plant be so produced, it should be most carefully guarded, multiplied by cuttings or grafts, and introduced far and wide. It is in this way that new varieties of fruits and flowers are produced from time to time.

Sometimes, too, a single bud on a tree will differ from the other buds and will produce a branch different from the other branches. This is known as bud variation. When there is thus developed a branch which happens to be of a superior kind, it should be propagated by cuttings just as you would propagate it if it had originated from a seed.

Mr. Gideon of Minnesota planted many apple seeds, and from them all raised one tree that was very fruitful, finely flavored, and able to withstand the cold Minnesota winter. This tree he multiplied by grafts and named the Wealthy apple. It is said that in giving this one apple to the world he benefited mankind to the value of more than one million dollars. It will be well to watch for any valuable bud or seed variant and never let a promising one be lost. Plants grown in this way from seeds are usually spoken of as seedlings.

PLANTS TO BE PROPAGATED FROM BUDS

The following list gives the names and methods by which our common garden fruits and flowers are propagated:

Figs: use cuttings 8 to 10 inches long or layer. Grapes: use long cuttings, layer, or graft upon old vines.

Apples: graft upon seedlings, usually crab seedlings one year old.

Pears: bud upon pear seedlings.

Cherries: bud upon cherry stock.

Plums: bud upon peach stock.

Peaches: bud upon peach or plum seedlings.

Quinces: use cuttings or layer.

Blackberries: propagate by suckers; cut from parent stem.

Black raspberries: layer; remove old stem.

Red raspberries: propagate by root-cuttings or suckers.

Strawberries: propagate by runners.

Currants and gooseberries: use long cuttings (these plants grow well only in cool climates; if attempted in warm climates, set in cold exposure).

Carnations, geraniums, roses, begonias, etc.: propagate by cuttings rooted in sand and then transplanted to small pots.

=EXERCISE=

Propagate fruits (grape, fig, strawberry) of various kinds; also ornamental plants. How long does it take them to root? Geraniums rooted in the spring will bloom in the fall. Do you know any one who selects seed potatoes properly? Make a careful selection of seed at the next harvest-time.

SECTION XVIII. PLANT SEEDING

In propagating by seed, as in reproducing by buds, we select a portion of the parent plant--for a seed is surely a part of the parent plant--and place it in the ground. There is, however, one great difference between a seed and a bud. The bud is really a piece of the parent plant, but a piece of one plant only, while a seed comes from the parts of two plants.

You will understand this fully if you read carefully Sections XIV-XVI. Since the seed is made of two plants, the plant that springs from a seed is much more likely to differ from its mother plant, that is, from the plant that produces the seed, than is a plant produced merely by buds. In some cases plants "come true to seed" very accurately. In others they vary greatly. For example, when we plant the seed of wheat, turnips, rye, onions, tomatoes, tobacco, or cotton, we get plants that are in most respects like the parent plant. On the other hand the seed of a Crawford peach or a Baldwin apple or a Bartlett pear will not produce plants like its parent, but will rather resemble its wild forefathers. These seedlings, thus taking after their ancestors, are always far inferior to our present cultivated forms. In such cases seeding is not practicable, and we must resort to bud propagation of one sort or another.

While in a few plants like those just mentioned the seed does not "come true," most plants, for example, cotton, tobacco, and others, do "come true." When we plant King cotton we may expect to raise King cotton. There will be, however, as every one knows, some or even considerable variation in the field. Some plants, even in exactly the same soil, will be better than the average, and some will be poorer. Now we see this variation in the plants of our field, and we believe that the plant will be in the main like its parent. What should we learn from this? Surely that if we wish to produce sturdy, healthy, productive plants we must go into our fields and pick out just such plants to secure seed from as we wish to produce another year. If we wait until the seed is separated from the plant that produced it before we select our cotton seed, we shall be planting seed from poor as well as from good plants, and must be content with a crop of just such stock as we have planted. By selecting seed from the most productive plants in the field and by repeating the selection each year, you can continually improve the breed of the plant you are raising. In selecting seed for cotton you may follow the plan suggested below for wheat.

The difference that you see between the wild and the cultivated chrysanthemums and between the samples of asparagus shown in Figs. 49 and 50 was brought about by just such continuous seed-selection from the kind of plant wanted.

By the careful selection of seed from the longest flax plants the increase in length shown in the accompanying figure was gained. The selection of seed from those plants bearing the most seed, regardless of the height of the plant, has produced flax like that to the right in the illustration. These two kinds of flax are from the same parent stock, but slight differences have been emphasized by continued seed-selection, until we now have really two varieties of flax, one a heavy seed-bearer, the other producing a long fiber.

You can in a similar way improve your cotton or any other seed crop. Sugar beets have been made by seed-selection to produce about double the percentage of sugar that they did a few years ago. Preparing and tilling land costs too much in money and work to allow the land to be planted with poor seed. When you are trying by seed-selection to increase the yield of cotton, there are two principles that should be borne in mind: first, seed should be chosen only from plants that bear many well-filled bolls of long-staple cotton; second, seed should be taken from no plant that does not by its healthy condition show hardihood in resisting disease and drouth.

The plan of choosing seeds from selected plants may be applied to wheat; but it would of course be too time-consuming to select enough single wheat plants to furnish all of the seed wheat for the next year. In this case adopt the following plan: In Fig. 52 let A represent the total size of your wheat field and let B represent a plat large enough to furnish seed for the whole field. At harvest-time go into section A and select the best plants you can find. Pick the heads of these and thresh them by hand. The seed so obtained must be carefully saved for your next sowing.

In the fall sow these selected seeds in area B. This area should produce the best wheat. At the next harvest cull not from the whole field but from the finest plants of plat B, and again save these as seed for plat B. Use the unculled seed from plat B to sow your crop. By following this plan continuously you will every year have seed from several generations of choice plants, and each year you will improve your seed.

It is of course advisable to move your seed plat B every year or two. For the new plat select land that has recently been planted in legumes. Always give this plat unwearying care.

In the selection of plants from which to get seed, you must know what kind of plants are really the best seed plants. First, you must not regard single heads or grains, but must select seed from the most perfect plant, looking at the plant as a whole and not at any single part of it. A first consideration is yield. Select the plants that yield best and are at the same time resistant to drouth, resistant to rust and to winter, early to ripen, plump of grain, and nonshattering. What a fine thing it would be to find even one plant free from rust in the midst of a rusted field! It would mean a rust-resistant plant. Its offspring also would probably be rust-resistant. If you should ever find such a plant, be sure to save its seed and plant it in a plat by itself. The next year again save seed from those plants least rusted. Possibly you can develop a rust-proof race of wheat! Keep your eyes open.

In England the average yield of wheat is thirty bushels an acre, in the United States it is less than fifteen bushels! In some states the yield is even less than nine bushels an acre. Let us select our seed with care, as the English people do, and then we can increase our yield. By careful seed-selection a plant-breeder in Minnesota increased the yield of his wheat by one fourth. Think what it would mean if twenty-five per cent were added to the world's supply of wheat at comparatively no cost; that is, at the mere cost of careful seed-selection. This would mean an addition to the world's income of about $500,000,000 each year. The United States would get about one fifth of this profit.

It often happens that a single plant in a crop of corn, cotton, or wheat will be far superior to all others in the field. Such a plant deserves special care. Do not use it merely as a seed plant, but carefully plant its seeds apart and tend carefully. The following season select the best of its offspring as favorites again. Repeat this selection and culture for several years until you fix the variety. This is the way new varieties are originated from plants propagated by seed.

In 1862 Mr. Abraham Fultz of Pennsylvania, while passing through a field of bearded wheat, found three heads of beardless, or bald, wheat. These he sowed by themselves that year, and as they turned out specially productive he continued to sow this new variety. Soon he had enough seed to distribute over the country. It became known as the Fultz wheat and is to-day one of the best varieties in the United States and in a number of foreign countries. Think how many bushels of wheat have been added to the world's annual supply by a few moments of intelligent observation and action on the part of this one man! He saw his opportunity and used it. How many similar opportunities do you think are lost? How much does your state or country lose thereby?

=EXERCISE=

Select one hundred seeds from a good, and one hundred from a poor, plant of the same variety. Sow them in two plats far enough apart to avoid cross-pollination, yet try to have soil conditions about the same. Give each the same care and compare the yield. Try this with corn, cotton, and wheat. Select seeds from the best plant in your good plat and from the poorest in your poor plat and repeat the experiment. This will require but a few feet of ground, and the good plat will pay for itself in yield, while the poor plat will more than pay in the lesson that it will teach you.

Write to the Department of Agriculture, Washington, D.C., and to your state experiment station for bulletins concerning seed-selection and methods of plant-improvement.

SECTION XIX. SELECTING SEED CORN

If a farmer would raise good crops he must, as already stated, select good seed. Many of the farmer's disappointments in the quantity and quality of his crops--disappointments often thought to come from other causes--are the result of planting poor seed. Seeds not fully ripened, if they grow at all, produce imperfect plants. Good seed, therefore, is the first thing necessary for a good crop. The seed of perfect plants only should be saved.

By wise and persistent selection, made in the field before the crop is fully matured, corn can be improved in size and made to mature earlier. Gather ears only from the most productive plants and save only the largest and best kernels.

You have no doubt seen the common American blackbirds that usually migrate and feed in such large numbers. They all look alike in every way. Now, has it ever occurred to you to ask why all blackbirds are black? The blackbirds are black simply because their parents are black.

Now in the same way that the young blackbirds resemble their parents, corn will resemble its parent stock. How many ears of corn do you find on a stalk? One, two, sometimes three or four. You find two ears of corn on a stalk because it is the nature of that particular stalk to produce two ears. In the same way the nature of some stalks is to produce but one ear, while it is the nature of others sometimes to produce two or more.

This resemblance of offspring to parent is known to scientists as heredity, or as "like producing like."

Some Southern corn-breeders take advantage of this law to improve their corn crop. If a stalk can be made to produce two ears of corn just as large as the single ear that most stalks bear, we shall get twice as much corn from a field in which the "two-eared" variety is planted. In the North and West the best varieties of corn have been selected to make but one ear to the stalk. It is generally believed that this is the best practice for the shorter growing seasons of the colder states.

These facts ought to be very helpful to us next year when our fathers are planting corn. We should get them to plant seed secured only from stalks that produced the most corn, whether the stalk had two or more ears or only one. If we follow this plan year by year, each acre of land will be made to produce more kernels and hence a larger crop of corn, and yet no more work will be required to raise the crop.

In addition to enlarging the yield of corn, you can, by proper selection of the best and most productive plants in the field, grow a new variety of seed corn. To do this you need only take the largest and best kernels from stalks bearing two ears; plant these, and at the next harvest again save the best kernels from stalks bearing the best ears. If you keep up this practice with great care for several years, you will get a vigorous, fruitful variety that will command a high price for seed.

=EXPERIMENT=

Every school boy and girl can make this experiment at leisure. From your own field get two ears of corn, one from a stalk bearing only one ear and the other from a stalk bearing two well-grown ears. Plant the grains from one ear in one plat, and the grains from the other in a plat of equal size. Use for both the same soil and the same fertilizer. Cultivate both plats in the same way. When the crop is ready to harvest, husk the corn, count the ears, and weigh the corn. Then write a short essay on your work and on the results and get your teacher to correct the story for your home paper.

SECTION XX. WEEDS

Have you ever noticed that some weeds are killed by one particular method, but that this same method may entirely fail to kill other kinds of weeds? If we wish to free our fields of weeds with the greatest ease, we must know the nature of each kind of weed and then attack it in the way in which we can most readily destroy it.

The ordinary pigweed (Fig. 56) differs from many other weeds in that it lives for only one year. When winter comes, it must die. Each plant, however, bears a great number of seeds. If we can prevent the plant from bearing seed in its first year, there will not be many seeds to come up the next season. In fact, only those seeds that were too deeply buried in the soil to come up the previous spring will be left, and of these two-year-old seeds many will not germinate. During the next season some old seeds will produce plants, but the number will be very much diminished. If care be exercised to prevent the pigweed from seeding again, and the same watchfulness be continued for a few seasons, this weed will be almost entirely driven from our fields.

A plant like the pigweed, which lives only one year, is called an annual and is one of the easiest weeds to destroy. Mustard, plantain, chess, dodder, cockle, crab grass, and Jimson weed are a few of our most disagreeable annual weeds.

The best time to kill any weed is when it is very small; therefore the ground in early spring should be constantly stirred in order to kill the young weeds before they grow to be strong and hardy.

The wild carrot differs from an annual in this way: it lives throughout one whole year without producing seeds. During its first year it accumulates a quantity of nourishment in the root, then rests in the winter. Throughout the following summer it uses this nourishment rapidly to produce its flowers and seeds. Then the plant dies. Plants that live through two seasons in this way are called biennials. Weeds of this kind may be destroyed by cutting the roots below the leaves with a grubbing-hoe or spud. A spud may be described as a chisel on a long handle (see Fig. 58). If biennials are not cut low enough they will branch out anew and make many seeds. Among the most common biennials are the thistle, moth mullein, wild carrot, wild parsnip, and burdock.

A third group of weeds consists of those that live for more than two years. These weeds are usually most difficult to kill. They propagate by means of running rootstocks as well as by seeds. Plants that live more than two seasons are known as perennials and include, for example, many grasses, dock, Canada thistle, poison ivy, passion flower, horse nettle, etc. There are many methods of destroying perennial weeds. They may be dug entirely out and removed. Sometimes in small areas they may be killed by crude sulphuric acid or may be starved by covering them with boards or a straw stack or in some other convenient way. A method that is very effective is to smother the weeds by a dense growth of some other plant, for example, cowpeas or buckwheat. Cowpeas are to be preferred, since they also enrich the soil by the nitrogen that the root-tubercles gather.

Weeds do injury in numerous ways; they shade the crop, steal its nourishment, and waste its moisture. Perhaps their only service is to make lazy people till their crops.

=EXERCISE=

You should learn to know by name the twenty worst weeds of your vicinity and to recognize their seeds. If there are any weeds you are not able to recognize, send a sample of each to your state experiment station. Make a collection, properly labeled, of weeds and weed seeds for your school.

SECTION XXI. SEED PURITY AND VITALITY

Seeds produce plants. The difference between a large and a small yield may depend upon the kind of plants we raise, and the kind of plant in turn is dependent upon the seeds that we sow.

Two things are important in the selection of seeds--purity and vitality. Seeds should be pure; that is, when sown they should produce no other plant than the one that we wish to raise. They should be able to grow. The ability of a seed to grow is termed its vitality. Good seed should be nearly or quite pure and should possess high vitality. The vitality of seeds is expressed as a per cent; for example, if 97 seeds out of 100 germinate, or sprout, the vitality is said to be 97. The older the seed the less is its vitality, except in a few rare instances in which seeds cannot germinate under two or three years.

Cucumber seeds may show 90 per cent vitality when they are one year old, 75 per cent when two years old, and 70 per cent when three years old--the per cent of vitality diminishing with increase of years. The average length of life of the seeds of cultivated plants is short: for example, the tomato lives four years; corn, two years; the onion, two years; the radish, five years. The cucumber seed may retain life after ten years; but the seeds of this plant too lose their vitality with an increase in years.

It is important when buying seeds to test them for purity and vitality. Dealers who are not honest often sell old seeds, although they know that seeds decrease in value with age. Sometimes, however, to cloak dishonesty they mix some new seeds with the old, or bleach old and yellow seeds in order to make them resemble fresh ones.

It is important, therefore, that all seeds bought of dealers should be thoroughly examined and tested; for if they do not grow, we not only pay for that which is useless but we are also in great danger of producing so few plants in our fields that we shall not get full use of the land, and thus we may suffer a more serious loss than merely paying for a few dead seeds. It will therefore be both interesting and profitable to learn how to test the vitality of seeds.

To test vitality plant one hundred seeds in a pot of earth or in damp sand, or place them between moist pieces of flannel, and take care to keep them moist and warm. Count those that germinate and thus determine the percentage of vitality. Germinating between flannel is much quicker than planting in earth. Care should be used to keep mice away from germinating seeds. (See Fig. 61.)

Sometimes the appearance of a package will show whether the seed has been kept in stock a long time. It is, however, much more difficult to find out whether the seeds are pure. You can of course easily distinguish seeds that differ much from those you wish to plant, but often certain weed seeds are so nearly like certain crop seeds as not to be easily recognized by the eye. Thus the dodder or "love vine," which so often ruins the clover crop, has seeds closely resembling clover seeds. The chess, or cheat, has seeds so nearly like oats that only a close observer can tell them apart. However, if you watch the seeds that you buy, and study the appearance of crop seeds, you may become expert in recognizing those that have no place in your planting.

One case is reported in which a seed-dealer intentionally allowed an impurity of 30 per cent to remain in the crop seeds, and this impurity was mainly of weed seeds. There were 450,000 of one kind and 288,000 of another in each pound of seed. Think of planting weeds at that rate! Sometimes three fourths of the seeds you buy are weed seeds.

In purchasing seeds the only safe plan is to buy of dealers whose reputation can be relied upon.

It not seldom happens that seeds, like corn, are stored in open cribs or barns before the moisture is entirely dried out of the seeds. Such seeds are liable to be frozen during a severe winter, and of course if this happens they will not sprout the following spring. The only way to tell whether such seeds have been killed is to test samples of them for vitality. Testing is easy; replanting is costly and often results in a short crop.

=EXERCISE=

Examine seeds both for vitality and purity. Write for farmers' bulletins on both these subjects. What would be the loss to a farmer who planted a ten-acre clover field with seeds that were 80 per cent bad? Can you recognize the seeds of the principal cultivated plants? Germinate some beet seeds. What per cent comes up? Can you explain? Collect for your school as many kinds of wild and cultivated seeds as you can.

CHAPTER IV

HOW TO RAISE A FRUIT TREE

Let each pupil grow an apple tree this year and attempt to make it the best in his neighborhood. In your attempt suppose you try the following plan. In the fall take the seed of an apple--a crab-apple is good--and keep it in a cool place during the winter. The simplest way to do this is to bury it in damp sand. In the spring plant it in a rich, loose soil.

Great care must be taken of the young shoot as soon as it appears above the ground. You want to make it grow as tall and as straight as possible during this first year of its life, hence you should give it rich soil and protect it from animals. Before the ground freezes in the fall take up the young tree with the soil that was around it and keep it all winter in a cool, damp place.

Now when spring comes it will not do to set out the carefully tended tree, for an apple tree from seed will not be a tree like its parent, but will tend to resemble a more distant ancestor. The distant ancestor that the young apple tree is most likely to take after is the wild apple, which is small, sour, and otherwise far inferior to the fruit we wish to grow. It makes little difference, therefore, what kind of apple seed we plant, since in any event we cannot be sure that the tree grown from it will bear fruit worth having unless we force it to do so.

SECTION XXII. GRAFTING

By a process known as grafting you can force your tree to produce whatever variety of apple you desire. Many people raise fruit trees directly from seed without grafting. Thus they often produce really worthless trees. By grafting they would make sure not only of having good trees rather than poor ones but also of having the particular kind of fruit that they wish. Hence you must now graft your tree.

First you must decide what variety of apple you want to grow on the tree. The Magnum Bonum is a great favorite as a fall apple. The Winesap is a good winter apple, while the Red Astrachan is a profitable early apple, especially in the lowland of the coast region. The Northern Spy, 苊 op, and Spitzenburg are also admirable kinds. Possibly some other apple that you know may suit your taste and needs better than any of these varieties.

If you have decided to raise an 苊 op or a Magnum Bonum or a Winesap, you must now cut a twig from the tree of your choice and graft it upon the little tree that you have raised. Choose a twig that is about the thickness of the young tree at the point where you wish to graft. Be careful to take the shoot from a vigorous, healthy part of the tree.

There are many ways in which you may join the chosen shoot or twig upon the young tree, but perhaps the best one for you to use is known as tongue grafting. This is illustrated in Fig. 64. The upper part, b, which is the shoot or twig that you cut from the tree, is known as the scion; the lower part, a, which is the original tree, is called the stock.

Cut the scion and stock as shown in Fig. 64. Join the cut end of the scion to the cut end of the stock. When you join them, notice that under the bark of each there is a thin layer of soft, juicy tissue. This is called the cambium. To make a successful graft the cambium in the scion must exactly join the cambium in the stock. Be careful, then, to see that cambium meets cambium. You now see why grafting can be more successfully done if you select a scion and stock of nearly the same size.

After fitting the parts closely together, bind them with cotton yarn (see Fig. 65) that has been coated with grafting wax. This wax is made of equal parts of tallow, beeswax, and linseed oil. Smear the wax thoroughly over the whole joint, and make sure that the joint is completely air-tight.

The best time to make this graft is when scion and stock are dormant, that is, when they are not in leaf. During the winter, say in February, is the best time to graft the tree. Set the grafted tree away again in damp sand until spring, then plant it in loose, rich soil.

Since all parts growing above the graft will be of the same kind as the scion, while all branches below it will be like the stock, it is well to graft low on the stock or even upon the root itself. The slanting double line in Fig. 66 shows the proper place to cut off for such grafting.

If you like you may sometime make the interesting and valuable experiment of grafting scions from various kinds of apple trees on the branches of one stock. In this way you can secure a tree bearing a number of kinds of fruit. You may thus raise the Bonum, Red Astrachan, Winesap, and as many other varieties of apples as you wish, upon one tree. For this experiment, however, you will find it better to resort to cleft grafting, which is illustrated in Fig. 68.

Luther Burbank, the originator of the Burbank potato, in attempting to find a variety of apple suited to the climate of California, grafted more than five hundred kinds of apple scions on one tree, so that he might watch them side by side and find out which kind was best suited to that state.

SECTION XXIII. BUDDING

If, instead of an apple tree, you were raising a plum or a peach tree, a form of propagation known as budding would be better than grafting. Occasionally budding is also employed for apples, pears, cherries, oranges, and lemons. Budding is done in the following manner. A single bud is cut from the scion and is then inserted under the bark of a one-year-old peach seedling, so that the cambium of the bud and stock may grow together.

Cut scions of the kind of fruit tree you desire from a one-year-old twig of the same variety. Wrap them in a clean, moist cloth until you are ready to use them. Just before using cut the bud from the scion, as shown in Fig. 69. This bud is now ready to be inserted on the north side of the stock, just two or three inches above the ground. The north side is selected to avoid the sun. Now, as shown at a in Fig. 70, make a cross and an up-and-down incision, or cut, on the stock; pull the bark back carefully, as shown in B; insert the bud C, as shown in D; then fold the bark back and wrap with yarn or raffia, as shown in E. As soon as the bud and branches have united, remove the wrapping to prevent its cutting the bark and cut the tree back close to the bud, as in Fig. 71, so as to force nourishment into the inserted bud.

Budding is done in the field without disturbing the tree as it stands in the ground. The best time to do budding is during the summer or fall months, when the bark is loose enough to allow the buds to be easily inserted.

Trees may be budded or grafted on one another only when they are nearly related. Thus the apple, crab-apple, hawthorn, and quince are all related closely enough to graft or bud on one another; the pear grows on some hawthorns, but not well on an apple; some chestnuts will unite with some kinds of oaks.

By using any of these methods you can succeed in getting with certainty the kind of tree that you desire.

SECTION XXIV. PLANTING AND PRUNING

The apple tree that you grafted should be set out in the spring. Dig a hole three or four feet in diameter where you wish the tree to grow. Place the tree in the hole and be very careful to preserve all the fine roots. Spread the roots out fully, water them, and pack fine, rich soil firmly about them. Place stakes about the young tree to protect it from injury. If the spot selected is in a windy location, incline the tree slightly toward the prevailing wind.

You must prune the tree as it grows. The object of pruning is to give the tree proper shape and to promote fruit-bearing. If the bud at the end of the main shoot grows, you will have a tall, cone-shaped tree. If, however, the end of the young tree be cut or "headed back" to the lines shown in Fig. 72, the buds below this point will be forced to grow and make a tree like that shown in Fig. 73. The proper height of heading for different fruits varies. For the apple tree a height of two or three feet is best.

Cutting an end bud of a shoot or branch always sends the nourishment and growth into the side buds. Trimming or pinching off the side buds throws the growth into the end bud. You can therefore cause your tree to take almost any shape you desire. The difference between the trees shown in Figs. 73 and 74 is entirely the result of pruning. Fig. 74 illustrates in general a correctly shaped tree. It is evenly balanced, admits light freely, and yet has enough foliage to prevent sun-scald. Figs. 75 and 76 show the effect of wisely thinning the branches.

The best time to prune is either in the winter or before the buds start in the spring. Winter pruning tends to favor wood-production, while summer pruning lessens wood-production and induces fruitage.

Each particular kind of fruit requires special pruning; for example, the peach should be made to assume the shape illustrated in Fig. 77. This is done by successive trimmings, following the plan illustrated in Figs. 71, 78, 79. You will gain several advantages from these trimmings. First, nourishment will be forced into the peach bud that you set on your stock. This will secure a vigorous growth of the scion. By a second trimming take off the "heel" (Fig. 78, h) close to the tree, and thus prevent decay at this point. One year after budding you should reduce the tree to a "whip," as in Fig. 79, by trimming at the dotted line in Fig. 78. This establishes the "head" of the tree, which in the case of the peach should be very low,--about sixteen inches from the ground,-- in order that a low foliage may lessen the danger of sun-scald to the main trunk.

In pruning never leave a stump such as is shown in Fig. 78, h. Such a stump, having no source of nourishment, will heal very slowly and with great danger of decay. If this heel is cleanly cut on the line ch (Fig. 78), the wound will heal rapidly and with little danger of decay. Leaving such a stump endangers the soundness of the whole tree. Fig. 80 shows the results of good and poor pruning on a large tree. When large limbs are removed it is best to paint the cut surface. The paint will ward off fungous disease and thus keep the tree from rotting where it was cut.

Pruning that leaves large limbs branching, as in Fig. 74, a, is not to be recommended, since the limbs when loaded with fruit or when beaten by heavy winds are liable to break. Decay is apt to set in at the point of breakage. The entrance of decay-fungi through some such wound or through a tiny crevice at such a crotch is the beginning of the end of many a fruitful tree.

Sometimes a tree will go too much to wood and too little to fruit. This often happens in rich soil and may be remedied by another kind of pruning known as root-pruning. This consists in cutting off a few of the roots in order to limit the food supply of the plant. You ought to learn more about root-pruning, however, before you attempt it.

How is a peach tree made? First, the blossom appears. Then pollination and fertilization occur. The fruit ripens. The pit, or seed, is saved. In the spring of the next year the seed is planted. The young tree, known as the stock, comes up quickly. In August of that year a bud of the variety which is wanted is inserted in the little stock, near the ground. One year later, in the spring, the stock is cut off just above the bud. The bud throws out a shoot, which grows to a height of about six feet, and in the fall this little peach tree is sold as a one-year-old tree. However, as is seen, the root is two years old.

How is an apple tree made? The seeds are saved in the fall of one year and planted the following year. The seedlings of the apple do not grow so rapidly as those of the peach. At the end of the year they are taken up and sorted, and in the following spring they are planted. In July or August they are budded. In the spring of the next year the stock is cut off above the bud, and the bud-shoot grows three or four feet. One year later the shoot branches and the top begins to form; and in the fall of the following year the tree may be sold as a two-year-old, although most persons prefer to buy it a year later as a three-year-old. In some parts of the country, particularly in the West, the little seedling is grafted in the second winter, in a grafting room, and the young grafts are set in the nursery row in the spring to complete their growth.

The planting in the orchard of the young peach and the young apple tree is done in practically the same way. After the hole for the tree has been dug and after proper soil has been provided, the roots should be spread and the soil carefully packed around them.

=EXERCISE=

Do you know any trees in your neighborhood that bear both wild and budded or grafted fruit? What are the chief varieties of apples grown in your neighborhood? grapes? currants? plums? cherries? figs? What is a good apple tree worth? Is there any land near by that could support a tree and is not now doing so? Examine several orchards and see whether the trees have the proper shape. Do you see any evidence of poor pruning? Do you find any heels? Can you see any place where heels have resulted in rotten or hollow trees? How could you have prevented this? Has the removal of branches ever resulted in serious decay? How is this to be prevented?

If your home is not well stocked with all the principal kinds of fruit, do you not want to propagate and attend to some of each kind? You will be surprised to find how quickly trees will bear and how soon you will be eating fruit from your own planting. Growing your own trees will make you feel proud of your skill.

CHAPTER V

HORTICULTURE

SECTION XXV. MARKET-GARDENING

The word horticulture is one of those broad words under which much is grouped. It includes the cultivation of orchard fruits, such as apples and plums; of small fruits, such as strawberries and raspberries; of garden vegetables for the table; of flowers of all sorts, including shrubbery and ornamental trees and their arrangement into beautiful landscape effects around our homes. Horticulture then is a name for an art that is both far-reaching and important.

The word gardening is generally given to that part of horticulture which has for its chief aim the raising of vegetables for our tables.

Flower-gardening, or the cultivation of plants valued for their bloom in making ornamental beds and borders and furnishing flowers for the decoration of the home, is generally called floriculture. Landscape-gardening is the art of so arranging flower-beds, grass, shrubbery, and trees as to produce pleasing effects in the grounds surrounding our homes and in great public parks and pleasure grounds.

Landscape-gardening, like architecture, has developed intoll as the artist makes them on canvas, but uses natural objects in his pictures instead of paint and canvas.

=Market-Gardening.= Formerly market-gardening was done on small tracts of land in the immediate vicinity of large cities, where supplies of stable manure could be used from the city stables. But with the great increase in the population of the cities, these small areas could no longer supply the demand, and the introduction of commercial fertilizers and the building of railroads enabled gardeners at great distances from city markets to grow and ship their products. Hence the markets, even in winter, are now supplied with fresh vegetables from regions where there is no frost. Then, as spring opens, fruits and vegetables are shipped from more temperate regions. Later vegetables and fruits come from the sections nearer the great cities. This gradual nearing of the supply fields continues until the gardens near the cities can furnish what is needed.

The market-gardeners around the great Northern cities, finding that winter products were coming from the South and from warmer regions, began to build hothouses and by means of steam and hot-water pipes to make warm climates in these glass houses. Many acres of land in the colder sections of the country are covered with heated glass houses, and in them during the winter are produced fine crops of tomatoes, lettuce, radishes, cauliflowers, eggplants, and other vegetables. The degree of perfection which these attain in spite of having such artificial culture, and their freshness as compared to the products brought from a great distance, have made winter gardening under glass a very profitable business. But it is a business that calls for the highest skill and the closest attention.

No garden, even for home use, is complete without some glass sashes, and the garden will be all the more successful if there is a small heated greenhouse for starting plants that are afterwards to be set in the garden.

=Hotbeds.= If there is no greenhouse, a hotbed is an important help in the garden. The bed is made by digging a pit two feet deep, seven feet wide, and as long as necessary.

The material for the hotbed is fresh horse manure mixed with leaves. This is thrown into a heap to heat. As soon as steam is seen coming from the heap the manure is turned over and piled again so that the outer part is thrown inside. When the whole is uniformly heated and has been turned two or three times, it is packed firmly into the pit already dug.

A frame six feet wide, twelve inches high on the north side and eight inches on the south side and as long as the bed is to be, is now made of plank. This is set upon the heated manure, thus leaving six inches on each side outside the frame. More manure is then banked all around it, and three or four inches of fine light and rich soil are placed inside the frame.

The frame is then covered with hotbed sashes six feet long and three feet wide. These slide up and down on strips of wood let into the sides of the frame. A thermometer is stuck into the soil and closely watched, for there will be too much heat at first for sowing seed. When the heat in the early morning is about 85? seeds may be sowed. The hotbed is used for starting tomato plants, eggplants, cabbage plants, and other vegetables that cannot stand exposure. It should be made about eight or ten weeks before the tender plants can be set out in the locality. In the South and Southwest it should be started earlier than in the North. For growing the best tomato plants, and for such hardy plants as lettuce and cabbage, it will be better to have cold-frames in addition to the hotbed; these need not be more than two or three sashes.

=Cold-Frames.= A cold-frame is like the frame used for a hotbed, but it is placed on well-manured soil in a sheltered spot. It is covered with the same kind of sashes and is used for hardening the plants sowed in the hotbed. The frame must be well banked with earth on the outside, and the glass must be covered on cold nights with straw, mats, or old carpets to keep out frost.

=Care of Hotbed and Cold-Frame.= If the sun be allowed to shine brightly on the glass of a cold-frame or hotbed, it will soon raise the temperature in the hotbed to a point that will destroy the plants. It is necessary, then, to pay close attention to the bed and, when the sun shines, to slip the sashes down or raise them and place a block under the upper end to allow the steam to pass off. The cold-frame also must be aired when the sun shines, and the sashes must be gradually slipped down in mild weather. Finally, they may be removed entirely on sunshiny days, so as to accustom the plants to the open air, but they must be replaced at night. For a while before setting the plants in the open gardens, leave the sashes off night and day.

While the hotbed may be used for starting plants, it is much better and more convenient to have a little greenhouse with fire heat for this purpose. A little house with but four sashes on each side will be enough to start a great many plants, and will also give room for some flowers in pots. With such a house a student can learn to manage a more extensive structure if he gives close attention to airing, watering, and keeping out insects.

=Sowing.= The time for sowing the different kinds of seeds is an important matter. Seeds vary greatly in their requirements. All need three conditions--a proper degree of heat, moisture, and air. Some seeds, like English peas, parsnips, beets, and radishes, will germinate and grow when the soil is still cool in the early spring, and peas will stand quite a frost after they are up. Therefore we plant English peas as early as the ground can be worked.

But if we should plant seeds like corn, string (or snap) beans, squashes, and other tender plants before the ground is warm enough, they would decay.

Seeds cannot germinate in soil that is perfectly dry, for there must be moisture to swell them and to start growth. The oxygen of the air is also necessary, and if seeds are buried so deeply that the air cannot reach them, they will not grow, even if they are warm and moist.

The depth of planting must vary with the character and size of the seed. English peas may be covered six inches deep and will be all the better for such covering, but if corn be covered so deep, it hardly gets above the ground. In planting small seeds like those of the radish, cabbage, turnip, lettuce, etc., a good rule is to cover them three times the thickness of the seed.

In sowing seeds when the ground is rather dry, it is a good plan, after covering them, to tramp on the row so as to press the soil closely to the seeds and to help it to retain moisture for germination, but do not pack the soil if it is damp.

In spring never dig or plow the garden while it is still wet, but always wait until the soil is dry enough to crumble freely.

=What Crops to grow.= The crops to be raised will of course depend upon each gardener's climate, surroundings, and markets. Sometimes it may pay a grower, if his soil and climate are particularly suited to one crop, to expend most of his time and energy on this crop; for example, in some sections of New York, on potatoes; in parts of Michigan, on celery; in Georgia, on watermelons; in western North Carolina, on cabbage. If circumstances allow this sort of gardening, it has many advantages, for of course it is much easier to acquire skill in growing one crop than in growing many.

On the other hand, it often happens that a gardener's situation requires him to grow most of the crops known to gardening. Each gardener then must be guided in his selection of crops by his surroundings.

=Care of Crops.= The gardener who wishes to attain the greatest success in his art must do four things:

First, he must make his land rich and keep it rich. Much of his success depends on getting his crops on the market ahead of other growers. To do this, his crops must grow rapidly, and crops grow rapidly only in rich soil. Then, too, land conveniently situated for market-gardening is nearly always costly. Hence the successful market-gardener must plan to secure the largest possible yield from as small an area as is practicable. The largest yield can of course be secured from the richest land.

Second, the gardener must cultivate his rich land most carefully and economically. He crowds his land with products that must grow apace. Therefore he, least of all growers, can afford to have any of his soil go to feed weeds, to have his land wash, or to have his growing crops suffer for lack of timely and wise cultivation. To cultivate his land economically the gardener must use the best tools and machines and the best methods of soil management.

Third, to get the best results he must grow perfect vegetables. To do this, he must add to good tillage a knowledge of the common plant diseases and of the ways of insects and bacterial pests; he must know how and when to spray, how and when to treat his seed, how and when to poison, how and when to trap his insect foes and to destroy their hiding-places.

Fourth, not only must the gardener grow perfect vegetables, but he must put them on the market in perfect condition and in attractive shape. Who cares to buy wilted, bruised, spoiling vegetables? Gathering, bundling, crating, and shipping are all to be watched carefully. Baskets should be neat and attractive, crates clean and snug, barrels well packed and well headed. Careful attention to all these details brings a rich return.

Among the gardener's important crops are the following:

=Asparagus.= This is a hardy plant. Its seed may be sowed either early in the spring or late in the fall. The seeds should be planted in rows. If the plants are well cultivated during the spring and summer, they will make vigorous roots for transplanting in the autumn.

In the fall prepare a piece of land by breaking it unusually deep and by manuring it heavily. After the land is thoroughly prepared, make in it furrows for the asparagus roots. These furrows should be six inches deep and three feet apart. Then remove the roots from the rows in which they have been growing during the summer, and set them two feet apart in the prepared furrows. Cover carefully at once.

In the following spring the young shoots must be well cultivated. In order to economize space, beets or lettuce may be grown between the asparagus rows during this first season. With the coming of cold weather the asparagus must again be freely manured and all dead tops cut off. Some plants will be ready for market the second spring. If the bed is kept free from weeds and well manured, it will increase in productiveness from year to year.

=Beans.= The most generally planted beans are those known as string, or snap, beans. Of the many varieties, all are sensitive to cold and hence must not be planted until frost is over.

Another widely grown kind of bean is the lima, or butter, bean. There are two varieties of the lima bean. One is large and generally grows on poles. This kind does best in the Northern states. The other is a small bean and may be grown without poles. This kind is best suited to the warmer climates of the Southern states.

=Cabbage.= In comparatively warm climates the first crop of cabbage is generally grown in the following way. The seeds are sowed in beds in September, and the plants grown from this sowing are in November transplanted to ground laid off in sharp ridges. The young plants are set on the south side of the ridges in order that they may be somewhat protected from the cold of winter. As spring comes on, the ridge is partly cut down at each working until the field is leveled, and thereafter the cultivation should be level.

Early cabbages need heavy applications of manure. In the spring, nitrate of soda applied in the rows is very helpful.

Seeds for the crop following this early crop should be sowed in March. Of course these seeds should be of a later variety than the first used. The young plants should be transplanted as soon as they are large enough. Early cabbages are set in rows three feet apart, the plants eighteen inches apart in the row. As the later varieties grow larger than the earlier ones, the plants should be set two feet apart in the row.

In growing late fall and winter cabbage the time of sowing varies with the climate. For the Northern and middle states, seeding should be done during the last of March and in April. South of a line passing west from Virginia it is hard to carry cabbages through the heat of summer and get them to head in the fall. However, if the seeds are sowed about the first of August in rich and moist soil and the plants set in the same sort of soil in September, large heads can be secured for the December market.

=Celery.= In the extreme northern part of our country, celery seeds are often sowed in a greenhouse or hotbed. This is done in order to secure plants early enough for summer blanching. This plan, however, suits only very cool climates.

In the middle states the seeds are usually sowed in a well-prepared bed about April. The young plants are moved to other beds as soon as they need room. Generally they are transplanted in July to rows prepared for them. These should be four feet apart, and the plants should be set six inches apart in the row. The celery bed should be carefully cultivated during the summer. In the fall, hill the stalks up enough to keep them erect. After the growing season is over dig them and set them in trenches. The trenches should be as deep as the celery is tall, and after the celery is put in them they should be covered with boards and straw.

In the more southern states, celery is usually grown in beds. The beds are generally made six feet wide, and rows a foot apart are run crosswise. The plants are set six inches apart, in September, and the whole bed is earthed up as the season advances. Finally, when winter comes the beds are covered with leaves or straw to prevent the plants from freezing. The celery is dug and bunched for market at any time during the winter.

By means of cold-frames a profitable crop of spring celery may be raised. Have the plants ready to go into the cold-frames late in October or early in November. The soil in the frame should be made very deep. The plants should make only a moderately rapid growth during the winter. In the early spring they will grow rapidly and so crowd one another as to blanch well. As celery grown in this way comes on the market at a time when no other celery can be had, it commands a good price.

In climates as warm as that of Florida, beds of celery can be raised in this way without the protection of cold-frames. A slight freeze does not hurt celery, but a long-continued freezing spell will destroy it.

Some kinds of celery seem to turn white naturally. These are called self-blanching kinds. Other kinds need to be banked with earth in order to make the stalks whiten. This kind usually gives the best and crispest stalks.

=Cucumbers and Cantaloupes.= Although cucumbers and cantaloupes are very different plants, they are grown in precisely the same way. Some gardeners plant them in hills. However, this is perhaps not the best plan. It is better to lay the land off in furrows six feet apart. After filling these with well-rotted stable manure, throw soil over them. Then make the top flat and plant the seeds. After the plants are up thin them out, leaving them a foot or more apart in the rows. Cultivate regularly and carefully until the vines cover the entire ground.

It is a good plan to sow cowpeas at the last working of cantaloupes, in order to furnish some shade for the melons. As both cucumbers and cantaloupes are easily hurt by cold, they should not be planted until the soil is warm and all danger of frost is past.

Cucumbers are always cut while they are green. They should never be pulled from the vine, but should always be cut with a piece of the stem attached. Cantaloupes should be gathered before they turn yellow and should be ripened in the house.

In some sections of the country the little striped cucumber-beetle attacks the melons and cucumbers as soon as they come up. These beetles are very active, and if their attacks are not prevented they will destroy the tender plants. Bone dust and tobacco dust applied just as the plants appear above the ground will prevent these attacks. This treatment not only keeps off the beetle, but also helps the growth of the plants.

=Eggplants.= Eggplants are so tender that they cannot be transplanted like tomatoes to cold-frames and gradually hardened to stand the cold spring air. These plants, started in a warm place, must be kept there until the soil to which they are to be transplanted is well warmed by the advance of spring. After the warm weather has fully set in, transplant them to rich soil, setting them three feet apart each way. This plant needs much manure. If large, perfect fruit is expected, the ground can hardly be made too rich.

Eggplants are subject to the same bacterial blight that is so destructive to tomatoes. The only way to prevent this disease is to plant in ground not lately used for tomatoes or potatoes.

=Onions.= The method of growing onions varies with the use to which it is intended to put them. To make the early sorts, which are eaten green in the spring, little onions called sets are planted. These are grown from seeds sowed late in the spring. The seeds are sowed thickly in rows in rather poor land. The object of selecting poor land is that the growth of the sets may be slow. When the sets have reached the size of small marbles, they are ready for the fall planting.

In the South the sets may be planted in September. Plant them in rows in rich and well-fertilized soil. They will be ready for market in March or April. In the more northerly states the sets are to be planted as early as possible in the spring.

To grow ripe onions the seeds must be sowed as early in the spring as the ground can be worked. The plants are thinned to a stand of three inches in the rows. As they grow, the soil is drawn away from them so that the onions sit on top of the soil with only their roots in the earth.

As soon as the tops ripen pull the onions and let them lie in the sun until the tops are dry. Then put them under shelter. As onions keep best with their tops attached, do not remove these until it is time for marketing.

=Peas.= The English pea is about the first vegetable of the season to be planted. It may be planted as soon as the ground is in workable condition. Peas are planted in rows, and it is a good plan to stretch wire netting for them to climb on. However, where peas are extensively cultivated they are allowed to fall on the ground.

There are many sorts of peas, differing both in quality and in time of production. The first to be planted are the extra-early varieties. These are not so fine as the later, wrinkled sorts, but the seeds are less apt to rot in cold ground. Following these, some of the fine, wrinkled sorts are to be planted in regular succession. Peas do not need much manure and do best in a light, warm soil.

=Tomatoes.= There is no vegetable grown that is more widely used than the tomato. Whether fresh or canned it is a staple article of food that can be served in many ways.

By careful selection and breeding, the fruit of the tomato has in recent years been much improved. There are now many varieties that produce perfectly smooth and solid fruit, and the grower can hardly go amiss in his selection of seeds if he bears his climate and his particular needs in mind.

Early tomatoes are started in the greenhouse or in the hotbed about ten weeks before the time for setting the plants in the open ground. They are transplanted to cold-frames as soon as they are large enough to handle. This is done to harden the plants and to give them room to grow strong before the final transplanting.

In kitchen gardens tomatoes are planted in rows four feet apart with the plants two feet apart in the rows. They are generally trained to stakes with but one stalk to a stake. When there is plenty of space, however, the plants are allowed to grow at will and to tumble on the ground. In this way they bear large crops. During the winter the markets are supplied with tomatoes either from tropical sections or from hothouses. As those grown in the hothouses are superior in flavor to those shipped from Florida and from the West Indies, and as they command good prices, great quantities are grown in this way.

In the South the bacterial blight which attacks the plants of this family is a serious drawback to tomato culture. The only way to escape this disease is to avoid planting tomatoes on land in which eggplants, tomatoes, or potatoes have been blighted. Lime spread around the plants seems to prevent the blight for one season on some soils.

At the approach of frost in the fall, green tomatoes can easily be preserved by wrapping them in paper. Gather them carefully and wrap each separately. Pack them in boxes and store in a cellar that is close enough to prevent the freezing of the fruit. A few days before the tomatoes are wanted for the table unpack as many as are needed, remove the paper, and allow them to ripen in a warm room.

Tomatoes require a rich soil. Scattering a small quantity of nitrate of soda around their roots promotes rapid growth.

=Watermelons.= As watermelons need more room than can usually be spared in a garden, they are commonly grown as a field crop.

A very light, sandy soil suits watermelons best. They can be grown on very poor soil if a good supply of compost be placed in each hill. The land for the melons should be laid off in about ten-foot checks; that is, the furrows should cross one another at right angles about every ten feet. A wide hole should be dug where the furrows cross, and into this composted manure should be put.

The best manure for watermelons is a compost of stable manure and wood-mold from the forest. Pile the manure and wood-mold in alternate layers for some time before the planting season. During the winter cut through the pile several times until the two are thoroughly mixed and finely pulverized. Be sure to keep the compost heap under shelter. Compost will lose in value if it is exposed to rains.

At planting-time, put two or three shovelfuls of this compost into each of the prepared holes, and over the top of the manure scatter a handful of any high-grade complete fertilizer. Then cover fertilizer and manure with soil, and plant the seeds in this soil. In cultivating, plow both ways of the checked rows and throw the earth toward the plants.

Some growers pinch off the vines when they have grown about three feet long. This is done to make them branch more freely, but the pinching is not necessary.

A serious disease, the watermelon wilt, is rapidly spreading through melon-growing sections. This disease is caused by germs in the soil, and the germs are hard to kill. If the wilt should appear in your neighborhood, do not allow any stable manure to be used on your melon land, for the germs are easily scattered by means of stable manure. The germs also cling to the seeds of diseased melons, and these seeds bear the disease to other fields. If you treat melon seeds as you are directed on page 135 to treat oat seeds, the germs on the seeds will be destroyed. By crossing the watermelon on the citron melon, a watermelon that is resistant to wilt has recently been developed and successfully grown in soils in which wilt is present. The new melon, inferior in flavor at first, is being improved from season to season and bids fair to rival other melons in flavor.

SECTION XXVI. FLOWER GARDENING

The comforts and joys of life depend largely upon small things. Of these small things perhaps none holds a position of greater importance in country life than the adornment of the home, indoors and outdoors, with flowers tastefully arranged. Their selection and planting furnish pleasant recreation; their care is a pleasing employment; and each little plant, as it sprouts and grows and develops, may become as much a pet as creatures of the sister animal kingdom. A beautiful, well-kept yard adds greatly to the pleasure and attractiveness of a country home. If a beautiful yard and home give joy to the mere passer-by, how much more must their beauty appeal to the owners. The decorating of the home shows ambition, pride, and energy--important elements in a successful life.

Plant trees and shrubs in your yard and border your masses of shrubbery with flower-beds. Do not disfigure a lawn by placing a bed of flowers in it. Use the flowers rather to decorate the shrubbery, and for borders along walks, and in the corners near steps, or against foundations.

If you wish to raise flowers for the sake of flowers, not as decorations, make the flower-beds in the back yard or at the side of the house.

Plants may be grown from seeds or from bulbs or from cuttings. The rooting of cuttings is an interesting task to all who are fond of flowers. Those who have no greenhouse and who wish to root cuttings of geraniums, roses, and other plants may do so in the following way. Take a shallow pan, an old-fashioned milk pan for instance, fill it nearly full of clean sand, and then wet the sand thoroughly. Stick the cuttings thickly into this wet sand, set the pan in a warm, sunny window, and keep the sand in the same water-soaked condition. Most cuttings will root well in a few weeks and may then be set into small flower-pots. Cuttings of tea roses should have two or three joints and be taken from a stem that has just made a flower. Allow one of the rose leaves to remain at the top of the cutting. Stick this cutting into the sand and it will root in about four weeks. Cuttings of Cape jasmine may be rooted in the same way. Some geraniums, the rose geranium for example, may be grown from cuttings of the roots.

Bulbs are simply the lower ends of the leaves of a plant wrapped tightly around one another and inclosing the bud that makes the future flower-stalk. The hyacinth, the narcissus, and the common garden onion are examples of bulbous plants. The flat part at the bottom of the bulb is the stem of the plant reduced to a flat disk, and between each two adjacent leaves on this flat stem there is a bud, just as above-ground there is a bud at the base of a leaf. These buds on the stem of the bulb rarely grow, however, unless forced to do so artificially. The number of bulbs may be greatly increased by making these buds grow and form other bulbs. In increasing hyacinths the matured bulbs are dug in the spring, and the under part of the flat stem is carefully scraped away to expose the base of the buds. The bulbs are then put in heaps and covered with sand. In a few weeks each bud has formed a little bulb. The gardener plants the whole together to grow one season, after which the little bulbs are separated and grown into full-sized bulbs for sale. Other bulbs, like the narcissus or the daffodil, form new bulbs that separate without being scraped.

There are some other plants which have underground parts that are commonly called bulbs but which are not bulbs at all; for example, the gladiolus and the caladium, or elephant's ear. Their underground parts are bulblike in shape, but are really solid flattened stems with eyes like the underground stem of the Irish potato. These parts are called corms. They may be cut into pieces like the potato and each part will grow.

The dahlia makes a mass of roots that look greatly like sweet potatoes, but there are no eyes on them as there are on the sweet potato. The only eyes are on the base of the stem to which they are joined. They may be sprouted like sweet potatoes and then soft cuttings made of the green shoots, after which they may be rooted in the greenhouse and later planted in pots.

There are many perennial plants that will bloom the first season when grown from the seed, though such seedlings are seldom so good as the plants from which they came. They are generally used to originate new varieties. Seeds of the dahlia, for instance, can be sowed in a box in a warm room in early March, potted as soon as the plants are large enough to handle, and finally planted in the garden when the weather is warm. They will bloom nearly as soon as plants grown by dividing the roots or from cuttings.

In growing annual plants from seed, there is little difficulty if the grower has a greenhouse or a hotbed with a glass sash. Even without these the plants may be grown in shallow boxes in a warm room. The best boxes are about four inches deep with bottoms made of slats nailed a quarter of an inch apart to give proper drainage. Some moss is laid over the bottom to prevent the soil from sifting through. The boxes should then be filled with light, rich soil. Fine black forest mold, thoroughly mixed with one fourth its bulk of well-rotted manure, makes the best soil for filling the seed-boxes. If this soil be placed in an oven and heated very hot, the heat will destroy many weeds that would otherwise give trouble. After the soil is put in the boxes it should be well packed by pressing it with a flat wooden block. Sow the seeds in straight rows, and at the ends of the rows put little wooden labels with the names of the flowers on them.

Seeds sowed in the same box should be of the same general size in order that they may be properly covered, for seeds need to be covered according to their size. After sowing the seed, sift the fine soil over the surface of the box. The best soil for covering small seeds is made by rubbing dry moss and leaf-mold through a sieve together. This makes a light cover that will not bake and will retain moisture. After covering the seeds, press the soil firm and smooth with a wooden block. Now sprinkle the covering soil lightly with a watering-pot until it is fairly moistened. Lay some panes of glass over the box to retain the moisture, and avoid further watering until moisture becomes absolutely necessary. Too much watering makes the soil too compact and rots the seed.

As soon as the seedlings have made a second pair of leaves, take them up with the point of a knife and transplant them into other boxes filled in the same way. They should be set two inches apart so as to give them room to grow strong. They may be transplanted from the boxes to the flower-garden by taking an old knife-blade and cutting the earth into squares, and then lifting the entire square with the plant and setting it where it is wanted.

There are many flower-seeds which are so small that they must not be covered at all. In this class we find begonias, petunias, and Chinese primroses. To sow these prepare boxes as for the other seeds, and press the earth smooth. Then scatter some fine, dry moss thinly over the surface of the soil. Sprinkle this with water until it is well moistened, and at once scatter the seeds thinly over the surface and cover the boxes with panes of glass until the seeds germinate. Transplant as soon as the young plants can be lifted out separately on the blade of a penknife.

Many kinds of flower-seeds may be sowed directly in the open ground where they are to remain. The sweet pea is one of the most popular flowers grown in this way. The seeds should be sowed rather thickly in rows and covered fully four inches deep. The sowing should be varied in time according to the climate. From North Carolina southward, sweet peas may be sowed in the fall or in January, as they are very hardy and should be forced to bloom before the weather becomes hot. Late spring sowing will not give fine flowers in the South. From North Carolina northward the seeds should be sowed just as early in the spring as the ground can be easily worked. When the plants appear, stakes should be set along the rows and a strip of woven-wire fence stretched for the plants to climb on. Morning-glory seeds are also sowed where they are to grow. The seeds of the moonflower are large and hard and will fail to grow unless they are slightly cut. To start their growth make a slight cut just through the hard outer coat of the seed so as to expose the white inside. In this way they will grow very readily. The seeds of the canna, or Indian-shot plant, are treated in a similar way to start them growing.

The canna makes large fleshy roots which in the North are taken up, covered with damp moss, and stored under the benches of the greenhouse or in a cellar. If allowed to get too dry, they will wither. From central North Carolina south it is best to cover them up thickly with dead leaves and let them stay in the ground where they grew. In the early spring take them up and divide for replanting.

Perennial plants, such as our flowering shrubs, are grown from cuttings of the ripe wood after the leaves have fallen in autumn. From North Carolina southward these cuttings should be set in rows in the fall. Cuttings ten inches long are set so that the tops are just even with the ground. A light cover of pine leaves will prevent damage from frost. Farther north the cuttings should be tied in bundles and well buried in the ground with earth heaped over them. In the spring set them in rows for rooting. In the South all the hardy hybrid perpetual roses can be grown in this way, and in any section the cuttings of most of the spring-flowering shrubs will grow in the same manner. The Japanese quince, which makes such a show of its scarlet flowers in early spring, can be best grown from three-inch cuttings made of the roots and planted in rows in the fall.

Many of our ornamental evergreen trees, such as the arbor vit? can be grown in the spring from seeds sowed in a frame. Cotton cloth should be stretched over the trees while they are young, to prevent the sun from scorching them. When a year old they may be set in nursery rows to develop until they are large enough to plant. Arbor vit?may also be grown from cuttings made by setting young tips in boxes of sand in the fall and keeping them warm and moist through the winter. Most of them will be rooted by spring.

The kinds of flowers that you can grow are almost countless. You can hardly make a mistake in selecting, as all are interesting. Start this year with a few and gradually increase the number under your care year by year, and aim always to make your plants the choicest of their kind.

Of annuals there are over four hundred kinds cultivated. You may select from the following list: phlox, petunias, China asters, California poppies, sweet peas, pinks, double and single sunflowers, hibiscus, candytuft, balsams, morning-glories, stocks, nasturtiums, verbenas, mignonette.

Of perennials select bleeding-hearts, pinks, bluebells, hollyhocks, perennial phlox, perennial hibiscus, wild asters, and goldenrods. From bulbs choose crocus, tulip, daffodil, narcissus, lily of the valley, and lily.

Some climbers are cob 鰈, honeysuckle, Virginia creeper, English ivy, Boston ivy, cypress vine, hyacinth bean, climbing nasturtiums, and roses.

To make your plants do best, cultivate them carefully. Allow no weeds to grow among them and do not let the surface of the soil dry into a hard crust. Beware, however, of stirring the soil too deep. Loosening the soil about the roots interrupts the feeding of the plant and does harm. Climbing plants may be trained to advantage on low woven-wire fences. These are especially serviceable for sweet peas and climbing nasturtiums. Do not let the plants go to seed, since seeding is a heavy drain on nourishment. Moreover, the plant has served its end when it seeds and is ready then to stop blossoming. You should therefore pick off the old flowers to prevent their developing seeds. This will cause many plants which would otherwise soon stop blossoming to continue bearing flowers for a longer period.

=Window-Gardening.= Growing plants indoors in the window possesses many of the attractions of outdoor flower-gardening, and is a means of beautifying the room at very small expense. Especially do window-gardens give delight during the barren winter time. They are a source of culture and pleasure to thousands who cannot afford extended and expensive ornamentation.

The window-garden may vary in size from an eggshell holding a minute plant to boxes filling all the available space about the window. The soil may be in pots for individual plants or groups of plants or in boxes for collections of plants. You may raise your flowers inside of the window on shelves or stands, or you may have a set of shelves built outside of the window and inclosed in glazed sashes. The illustration on page 119 gives an idea of such an external window-garden.

The soil must be rich and loose. The best contains some undecayed organic matter such as leaf-mold or partly decayed sods and some sand. Raise your plants from bulbs, cuttings, or seed, just as in outdoor gardens. Some plants do better in cool rooms, others in a warmer temperature.

If the temperature ranges from 35?to 70? averaging about 55? azaleas, daisies, carnations, candytuft, alyssum, dusty miller, chrysanthemums, cinerarias, camellias, daphnes, geraniums, petunias, violets, primroses, and verbenas make especially good growths.

If the temperature is from 50?to 90? averaging 70? try abutilon, begonia, bouvardia, caladium, canna, Cape jasmine, coleus, fuchsia, gloxinia, heliotrope, lantana, lobelia, roses, and smilax.

If your box or window is shaded a good part of the time, raise begonias, camellias, ferns, and Asparagus Sprengeri.

[Illustration: FIG. 111. FERNS FOR BOTH INDOORS AND OUTDOORS]

When the soil is dry, water it; then apply no more water until it again becomes dry. Beware of too much water. The plants should be washed occasionally with soapsuds and then rinsed. If red spiders are present, sponge them off with water as hot as can be borne comfortably by the hand. Newspapers afford a good means of keeping off the cold.

CHAPTER VI

THE DISEASES OF PLANTS

SECTION XXVII. THE CAUSE AND NATURE OF PLANT DISEASE

Plants have diseases just as animals do; not the same diseases, to be sure, but just as serious for the plant. Some of them are so dangerous that they kill the plant; others partly or wholly destroy its usefulness or its beauty. Some diseases are found oftenest on very young plants, others prey on the middle-aged tree, while still others attack merely the fruit. Whenever a farmer or fruit-grower has disease on his plants, he is sure to lose much profit.

You have all seen rotten fruit. This is diseased fruit. Fruit rot is a plant disease. It costs farmers millions of dollars annually. A fruit-grower recently lost sixty carloads of peaches in a single year through rot which could have been largely prevented if he had known how.

Many of the yellowish or discolored spots on leaves are the result of disease, as is also the smut of wheat, corn, and oats, the blight of the pear, and the wilt of cotton. Many of these diseases are contagious, or, as we often hear said of measles, "catching." This is true, among others, of the apple and peach rots. A healthy apple can catch this disease from a sick apple. You often see evidence of this in the apple bin. So, too, many of the diseases found in the field or garden are contagious.

Sometimes when the skin of a rotten apple has been broken you will find in the broken place a blue mold. It was this that caused the apple to decay. This mold is a living plant; very small, certainly, but nevertheless a plant. Let us learn a little about molds, in order that we may better understand our apple and potato rots, as well as other plant diseases.

If you cut a lemon and let it stand for a day or two, there will probably appear a blue mold like that you have seen on the surface of canned fruit. Bread also sometimes has this blue mold; at other times bread has a black mold, and yet again a pink or a yellow mold.

These and all other molds are tiny living plants. Instead of seeds they produce many very small bodies that serve the purpose of seeds and reproduce the mold. These are called spores. Fig. 112 shows how they are borne on the parent plant.

It is also of great importance to decide whether by keeping the spores away we may prevent mold. Possibly this experiment will help us. Moisten a piece of bread, then dip a match or a pin into the blue mold on a lemon, and draw the match across the moist bread. You will thus plant the spores in a row, though they are so small that perhaps you may not see any of them. Place the bread in a damp place for a few days and watch it. Does the mold grow where you planted it? Does it grow elsewhere? This experiment should prove to you that molds are living things and can be planted. If you find spots elsewhere, you must bear in mind that these spores are very small and light and that some of them were probably blown about when you made your sowing. When you touch the moldy portion of a dry lemon, you see a cloud of dust rise. This dust is made of millions of spores.

If you plant many other kinds of mold you will find that the molds come true to the kind that is planted; that like produces like even among molds.

You can prove, also, that the mold is caused only by other mold. To do this, put some wet bread in a wide-mouthed bottle and plug the mouth of the bottle with cotton. Kill all the spores that may be in this bottle by steaming it an hour in a cooking-steamer. This bread will not mold until you allow live mold from the outside to enter. If, however, at any time you open the bottle and allow spores to enter, or if you plant spores therein, and if there be moisture enough, mold will immediately set in.

The little plants which make up these molds are called fungi. Some fungi, such as the toadstools, puffballs, and devil's snuff-box, are quite large; others, namely the molds, are very small; and others are even smaller than the molds. Fungi never have the green color of ordinary plants, always reproduce by spores, and feed on living matter or matter that was once alive. Puffballs, for example, are found on rotting wood or dead twigs or roots. Some fungi grow on living plants, and these produce plant disease by taking their nourishment from the plant on which they grow; the latter plant is called the host.

The same blue mold that grows on bread often attacks apples that have been slightly bruised; it cannot pierce healthy apple skin. You can plant the mold in the bruised apple just as you did on bread and watch its rapid spread through the apple. You learn from this the need of preventing bruised or decayed apples from coming in contact with healthy fruit.

Just as the fungus studied above lives in the apple or bread, so other varieties live on leaves, bark, etc. Fig. 113 represents the surface of a mildewed rose leaf greatly magnified. This mildew is a fungus. You can see its creeping stems, its upright stalk, and numerous spores ready to fall off and spread the disease with the first breath of wind. You must remember that this figure is greatly magnified, and that the whole portion shown in the figure is only about one tenth of an inch across. Fig. 114 shows the general appearance of a twig affected by this disease.

Mildew on the rose or on any other plant may be killed by spraying the leaves with a solution of liver of sulphur; to make this solution, use one ounce of the liver of sulphur to two gallons of water.

The fungus that causes the pear-leaf spots has its spores in little pits (Fig. 115). The spores of some fungi also grow on stalks, as shown in Fig. 116. This figure represents an enlarged view of the pear scab, which causes so much destruction.

You see, then, that fungi are living plants that grow at the expense of other plants and cause disease. Now if you can cover the leaf with a poison that will kill the spore when it comes, you can prevent the disease. One such poison is the Bordeaux mixture, which has proved of great value to farmers.

Since the fungus in most cases lives within the leaves, the poison on the outside does no good after the fungus is established. The treatment can be used only to prevent attack, not to cure, except in the case of a few mildews that live on the outside of the leaf, as does the rose mildew.

=EXERCISE=

Why do things mold more readily in damp places? Do you now understand why fruit is heated before it is canned? Try to grow several kinds of mold. Do you know any fungi which may be eaten?

Transfer disease from a rotten apple to a healthy one and note the rapidity of decay. How many really healthy leaves can you find on a strawberry plant? Do you find any spots with reddish borders and white centers? Do you know that this is a serious disease of the strawberry? What damage does fruit mold do to peaches, plums, or strawberries?

Write to your experiment station for bulletins on plant diseases and methods for making and using spraying mixtures.

SECTION XXVIII. YEAST AND BACTERIA

Can you imagine a plant so small that it would take one hundred plants lying side by side to equal the thickness of a sheet of writing-paper? There are plants that are so small. Moreover, these same plants are of the utmost importance to man. Some of them do him great injury, while others aid him very much.

You will see their importance when you are told that certain of them in their habits of life cause great change in the substances in which they live. For example, when living in a sugary substance they change the sugar into a gas and an alcohol. Do you remember the bright bubbles of gas you have seen rising in sweet cider or in wine as it soured? These bubbles are caused by one of these small plants--the yeast plant. As the yeast plant grows in the sweet fruit juice, alcohol is made and a gas is given off at the same time, and this gas makes the bubbles.

Later, other kinds of plants equally small will grow and change the alcohol into an acid which you will recognize as vinegar by its sour taste and peculiar odor. Thus vinegar is made by the action of two different kinds of little living plants in the cider. That these are living beings you can prove by heating the cider and keeping it tightly sealed so that nothing can enter it. You will find that because the living germs have been killed by the heat, the cider will not ferment or sour as it did before. The germs could of course be killed by poisons, but then the cider would be unfit for use. It is this same little yeast plant that causes bread to rise.

When you see any decaying matter you may know that in it minute plants much like the yeast plant are at work. Since decay is due to them, we take advantage of the fact that they cannot grow in strong brine or smoke; and we prepare meat for keeping by salting it or by smoking it or by both of these methods.

You see that some of the yeast plants and bacteria, as many of these forms are called, are very friendly to us, while others do us great harm.

Some bacteria grow within the bodies of men and other animals or in plants. When they do so they may produce disease. Typhoid fever, diphtheria, consumption, and many other serious diseases are caused by bacteria. Fig. 118, e, shows the bacterium that causes typhoid fever. In the picture, of course, it is very greatly magnified. In reality these bacteria are so small that about twenty-five thousand of them side by side would extend only one inch. These small beings produce their great effects by very rapid multiplication and by giving off powerful poisons.

Bacteria are so small that they are readily borne on the dust particles of the air and are often taken into the body through the breath and also through water or milk. You can therefore see how careful you should be to prevent germs from getting into the air or into water or milk when there is disease about your home. You should heed carefully all instructions of your physician on this point, so that you may not spread disease.

SECTION XXIX. PREVENTION OF PLANT DISEASE

In the last two sections you have learned something of the nature of those fungi and bacteria that cause disease in animals and plants. Now let us see how we can use this knowledge to lessen the diseases of our crops. Farmers lose through plant diseases much that could be saved by proper precaution.

First, you must remember that every diseased fruit, twig, or leaf bears millions of spores. These must be destroyed by burning. They must not be allowed to lie about and spread the disease in the spring. See that decayed fruit in the bin or on the trees is destroyed in the same manner. Never throw decayed fruit into the garden or orchard, as it may cause disease the following year.

Second, you can often kill spores on seeds before they are planted and thus prevent the development of the fungus (see pp. 134-137).

Third, often the foliage of the plant can be sprayed with a poison that will prevent the germination of the spores (see pp. 138-140).

Fourth, some varieties of plants resist disease much more stoutly than others. We may often select the resistant form to great advantage (see Fig. 119).

Fifth, after big limbs are pruned off, decay often sets in at the wound. This decay may be prevented by coating the cut surface with paint, tar, or some other substance that will not allow spores to enter the wound or to germinate there.

Sixth, it frequently happens that the spore or fungus remains in the soil. This is true in the cotton wilt, and the remedy is so to rotate crops that the diseased land is not used again for this crop until the spores or fungi have died.

SECTION XXX. SOME SPECIAL PLANT DISEASES

=Fire-Blight of the Pear and Apple.= You have perhaps heard your father speak of the "fire-blight" of pear and apple trees. This is one of the most injurious and most widely known of fruit diseases. Do you want to know the cause of this disease and how to prevent it?

First, how will you recognize this disease? If the diseased bough at which you are looking has true fire-blight, you will see a blackened twig with withered, blackened leaves. During winter the leaves do not fall from blighted twigs as they do from healthy ones. The leaves wither because of the diseased twig, not because they are themselves diseased. Only rarely does the blight really enter the leaf. Sometimes a sharp line separates the blighted from the healthy part of the twig.

This disease is caused by bacteria, of which you have read in another section. The fire-blight bacteria grow in the juicy part of the stem, between the wood and the bark. This tender, fresh layer (as explained on page 79) is called the cambium, and is the part that breaks away and allows you to slip the bark off when you make your bark whistle in the spring. The growth of new wood takes place in the cambium, and this part of the twig is therefore full of nourishment. If this nourishment is stolen the plant of course soon suffers.

The bacteria causing fire-blight are readily carried from flower to flower and from twig to twig by insects; therefore to keep these and other bacteria away from your trees you must see to it that all the trees in the neighborhood of your orchard are kept free from mischievous enemies. If harmful bacteria exist in near-by trees, insects will carry them to your orchard. You must therefore watch all the relatives of the pear; namely, the apple, hawthorn, crab, quince, and mountain ash, for any of these trees may harbor the germs.

When any tree shows blight, every diseased twig on it must be cut off and burned in order to kill the germs, and you must cut low enough on the twig to get all the bacteria. It is best to cut a foot below the blackened portion. If by chance your knife should cut into wood containing the living germs, and then you should cut into healthy wood with the same knife, you yourself would spread the disease. It is therefore best after each cutting to dip your knife into a solution of carbolic acid. This will kill all bacteria clinging to the knife-blade. The surest time to do complete trimming is after the leaves fall in the autumn, as diseased twigs are most easily recognized at that time, but the orchard should be carefully watched in the spring also. If a large limb shows the blight, it is perhaps best to cut the tree entirely down. There is little hope for such a tree.

A large pear-grower once said that no man with a sharp knife need fear the fire-blight. Yet our country loses greatly by this disease each year.

It may be added that winter pruning tends to make the tree form much new wood and thus favors the disease. Rich soil and fertilizers make it much easier in a similar way for the tree to become a prey to blight.

=EXERCISE=

Ask your teacher to show you a case of fire-blight on a pear or apple tree. Can you distinguish between healthy and diseased wood? Cut the twig open lengthwise and see how deep into the wood and how far down the stem the disease extends. Can you tell surely from the outside how far the twig is diseased? Can you find any twig that does not show a distinct line of separation between diseased and healthy wood? If so, the bacteria are still living in the cambium. Cut out a small bit of the diseased portion and insert it under the bark of a healthy, juicy twig within a few inches of its tip and watch it from day to day. Does the tree catch the disease? This experiment may prove to you how easily the disease spreads. If you should see any drops like dew hanging from diseased twigs, touch a little of this moisture to a healthy flower and watch for results.

Cut and burn all diseased twigs that you can find. Estimate the damage done by fire-blight.

Farmers' bulletins on orchard enemies are published by the Department of Agriculture, Washington, D.C., and can be had by writing for them. They will help your father much in treating fire-blight.

=Oat Smuts.= Let us go out into a near-by oat field and look for all the blackened heads of grain that we can find. How many are there? To count accurately let us select an area one foot square. We must look carefully, for many of these blackened heads are so low that we shall not see them at the first glance. You will be surprised to find as many as thirty or forty heads in every hundred so blackened. These blackened heads are due to a plant disease called smut.

When threshing-time comes you will notice a great quantity of black dust coming from the grain as it passes through the machine. The air is full of it. This black dust consists of the spores of a tiny fungous plant. The fungous smut plant grows upon the oat plant, ripens its spores in the head, and is ready to be thoroughly scattered among the grains of the oats as they come from the threshing-machine.

These spores cling to the grain and at the next planting are ready to attack the sprouting plantlet. A curious thing about the smut is that it can gain a foothold only on very young oat plants; that is, on plants about an inch long or of the age shown in Fig. 121.

When grain covered with smut spores is planted, the spores develop with the sprouting seeds and are ready to attack the young plant as it breaks through the seed-coat. You see, then, how important it is to have seed grain free from smut. A substance has been found that will, without injuring the seeds, kill all the smut spores clinging to the grain. This substance is called formalin. Enough seed to plant a whole acre can be treated with formalin at a cost of only a few cents. Such treatment insures a full crop and clean seed for future planting. Try it if you have any smut.

Fig. 122 illustrates what may be gained by using seeds treated to prevent smut. The annual loss to the farmers of the United States from smut on oats amounts to several millions of dollars. All that is needed to prevent this loss is a little care in the treatment of seed and a proper rotation of crops.

=EXERCISE=

Count the smutted heads on a patch three feet square and estimate the percentage of smut in all the wheat and oat fields near your home. On which is it most abundant? Do you know of any fields that have been treated for smut? If so, look for smut in these fields. Ask how they were treated. Do you know of any one who uses bluestone for wheat smut? Can oats be treated with bluestone?

At planting time get an ounce of formalin at your drug store or from the state experiment station. Mix this with three gallons of water. This amount will treat three bushels of seeds. Spread the seeds thinly on the barn floor and sprinkle them with the mixture, being careful that all the seeds are thoroughly moistened. Cover closely with blankets for a few hours and plant very soon after treatment. Try this and estimate the per cent of smut at next harvest-time. Write to your experiment station for a bulletin on smut treatment.

=Potato Scab.= The scab of the white, or Irish, potato is one of the commonest and at the same time most easily prevented of plant diseases. Yet this disease diminishes the profits of the potato-grower very materially. Fig. 123 shows a very scabby potato, while Fig. 124 represents a healthy one. This scab is caused by a fungous growth on the surface of the potato. Of course it lessens the selling-price of the potatoes. If seed potatoes be treated to a bath of formalin just before they are planted, the formalin will kill the fungi on the potatoes and greatly diminish the amount of scab at the next harvest. Therefore before they are planted, seed potatoes should be soaked in a weak solution of formalin for about two hours. One-half pint of formalin to fifteen gallons of water makes a proper solution.

One pint of formalin, or enough for thirty gallons of water, will cost but thirty-five cents. Since this solution can be used repeatedly, it will do for many bushels of seed potatoes.

=Late Potato Blight.= The blight is another serious disease of the potato. This is quite a different disease from the scab and so requires different treatment. The blight is caused by another fungus, which attacks the foliage of the potato plant. When the blight seriously attacks a crop, it generally destroys the crop completely. In the year 1845 a potato famine extending over all the United States and Europe was caused by this disease.

Spraying is the remedy for potato blight. Fig. 128 shows the effect of spraying upon the yield. In this case the sprayed field yielded three hundred and twenty-four bushels an acre, while the unsprayed yielded only one hundred bushels to an acre. Fig. 127 shows the result of three applications of the spraying mixture on the diseased field. Figs. 129 and 130 show how the spraying is done.

=EXERCISE=

Watch the potatoes at the next harvest and estimate the number that is damaged by scab. You will remember that formalin is the substance used to prevent grain smuts. Write to your state experiment station for a bulletin telling how to use formalin, as well as for information regarding other potato diseases. Give the treatment a fair trial in a portion of your field this year and watch carefully for results. Make an estimate of the cost of treatment and of the profits. How does the scab injure the value of the potato? The late blight can often be recognized by its odor. Did you ever smell it as you passed an affected field?

=Club Root.= Club root is a disease of the cabbage, turnip, cauliflower, etc. Its general effect is shown in the illustration (Fig. 131). Sometimes this disease does great damage. It can be prevented by using from eighty to ninety bushels of lime to an acre.

=Black Knot.= Black knot is a serious disease of the plum and of the cherry tree. It attacks the branches of the tree; it is well illustrated in Fig. 132. Since it is a contagious disease, great care should be exercised to destroy all diseased branches of either wild or cultivated plums or cherries. In many states its destruction is enforced by law. All black knot should be cut out and burned some time before February of each year. This will cost little and save much.

=Peach Leaf Curl.= Peach leaf curl does damage amounting to about $3,000,000 yearly in the United States. It can be almost entirely prevented by spraying the tree with Bordeaux mixture or lime-sulphur wash before the buds open in the spring. It is not safe to use strong Bordeaux mixture on peach trees when they are in leaf.

=Cotton Wilt.= Cotton wilt when it once establishes itself in the soil completely destroys the crop. The fungus remains in the soil, and no amount of spraying will kill it. The only known remedy is to cultivate a resistant variety of cotton or to rotate the crop.

=Fruit Mold.= Fruit mold, or brown rot, often attacks the unripe fruit on the tree, and turns it soft and brown and finally fuzzy with a coat of mildew. Fig. 133 shows some peaches thus attacked. Often the fruits do not fall from the trees but shrivel up and become "mummies" (Fig. 134). This rot is one of the most serious diseases of plums and peaches. It probably diminishes the value of the peach harvest from 50 to 75 per cent. Spraying according to the directions in the Appendix will kill the disease.

CHAPTER VII

ORCHARD, GARDEN, AND FIELD INSECTS

SECTION XXXI. INSECTS IN GENERAL

The farmer who has fought "bugs" on crop after crop needs no argument to convince him that insects are serious enemies to agriculture. Yet even he may be surprised to learn that the damage done by them, as estimated by good authority, amounts to millions and millions of dollars yearly in the United States and Canada.

Every one thinks he knows what an insect is. If, however, we are willing in this matter to make our notion agree with that of the people who have studied insects most and know them best, we must include among the true insects only such air-breathing animals as have six legs, no more, and have the body divided into three parts--head, thorax, and abdomen. These parts are clearly shown in Fig. 136, which represents the ant, a true insect. All insects do not show the divisions of the body so clearly as this figure shows them, but on careful examination you can usually make them out. The head bears one pair of feelers, and these in many insects serve also as organs of smell and sometimes of hearing. Less prominent feelers are to be found in the region of the mouth. These serve as organs of taste.

The eyes of insects are especially noticeable. Close examination shows them to be made up of a thousand or more simple eyes. Such an eye is called a compound eye. An enlarged view of one of these is shown in Fig. 138.

Attached to the thorax are the legs and also the wings, if the insect has wings. The rear portion is the abdomen, and this, like the other parts, is composed of parts known as segments. The insect breathes through openings in the abdomen and thorax called spiracles (see Fig. 137).

An examination of spiders, mites, and ticks shows eight legs; therefore these do not belong to the true insects, nor do the thousand-legged worms and their relatives.

The chief classes of insects are as follows: the flies, with two wings only; the bees, wasps, and ants, with four delicate wings; the beetles, with four wings--two hard, horny ones covering the two more delicate ones. When the beetle is at rest its two hard wings meet in a straight line down the back. This peculiarity distinguishes it from the true bug, which has four wings. The two outer wings are partly horny, and in folding lap over each other. Butterflies and moths are much alike in appearance but differ in habit. The butterfly works by day and the moth by night. Note the knob on the end of the butterfly's feeler (Fig. 143). The moth has no such knob.

It is important to know how insects take their food, for by knowing this we are often able to destroy insect pests. Some are provided with mouth parts for chewing their food; others have a long tube with which they pierce plants or animals and, like the mosquito, suck their food from the inside. Insects of this latter class cannot of course be harmed by poison on the surface of the leaves on which they feed.

Many insects change their form from youth to old age so much that you can scarcely recognize them as the same creatures. First comes the egg. The egg hatches into a worm-like animal known as a grub, maggot, or caterpillar, or, as scientists call it, a larva. This creature feeds and grows until finally it settles down and spins a home of silk, called a cocoon (Fig. 145). If we open the cocoon we shall find that the animal is now covered with a hard outside skeleton, that it cannot move freely, and that it cannot eat at all. The animal in this state is known as the pupa (Figs. 145 and 146). Sometimes, however, the pupa is not covered by a cocoon, sometimes it is soft, and sometimes it has some power of motion (Fig. 141). After a rest in the pupa stage the animal comes out a mature insect (Figs. 142 and 143).

From this you can see that it is especially important to know all you can about the life of injurious insects, since it is often easier to kill these pests at one stage of their life than at another. Often it is better to aim at destroying the seemingly harmless beetle or butterfly than to try to destroy the larv?that hatch from its eggs, although, as you must remember, it is generally the larv?that do the most harm. Larv?grow very rapidly; therefore the food supply must be great to meet the needs of the insect.

Some insects, the grasshopper for example, do not completely change their form. Fig. 147 represents some young grasshoppers, which very closely resemble their parents.

Insects lay many eggs and reproduce with remarkable rapidity. Their number therefore makes them a foe to be much dreaded. The queen honeybee often lays as many as 4000 eggs in twenty-four hours. A single house fly lays between 100 and 150 eggs in one day. The mosquito lays eggs in quantities of from 200 to 400. The white ant often lays 80,000 in a day, and so continues for two years, probably laying no less than 40,000,000 eggs. In one summer the bluebottle fly could have 500,000,000 descendants if they all lived. The plant louse, at the end of the fifth brood, has laid in a single year enough eggs to produce 300,000,000 young. Of course every one knows that, owing to enemies and diseases (for the insects have enemies which prey on them just as they prey on plants) comparatively few of the insects hatched from these eggs live till they are grown.

The number of insects which are hurtful to crops, gardens, flowers, and forests seems to be increasing each season. Therefore farm boys and girls should learn to recognize these harmful insects and to know how they live and how they may be destroyed. Those who know the forms and habits of these enemies of plants and trees are far better prepared to fight them than are those who strike in the dark. Moreover such knowledge is always a source of interest and pleasure. If you begin to study insects, you will soon find your love for the study growing.

=EXERCISE=

Collect cocoons and pup?of insects and hatch them in a breeding-cage similar to the one illustrated in Fig. 149. Make several cages of this kind. Collect larv?of several kinds; supply them with food from plants upon which you found them. Find out the time it takes them to change into another stage. Write a description of this process.

The plant louse could produce in its twelfth brood 10,000,000,000,000,000,000,000 offspring. Each louse is about one tenth of an inch long. If all should live and be arranged in single file, how many miles long would such a procession be?

SECTION XXXII. ORCHARD INSECTS

=The San Jose Scale.= The San Jose scale is one of the most dreaded enemies of fruit trees. It is in fact an outlaw in many states. It is an unlawful act to sell fruit trees affected by it. Fig. 150 shows a view of a branch nearly covered with this pest. Although this scale is a very minute animal, yet so rapidly does it multiply that it is very dangerous to the tree. Never allow new trees to be brought into your orchard until you feel certain that they are free from the San Jose scale. If, however, it should in any way gain access to your orchard, you can prevent its spreading by thorough spraying with what is known as the lime-sulphur mixture. This mixture has long been used on the Pacific coast as a remedy for various scale insects. When it was first tried in other parts of the United States the results were not satisfactory and its use was abandoned. However, later experiments with it have proved that the mixture is thoroughly effective in killing this scale and that it is perfectly harmless to the trees. Until the lime-sulphur mixture proved to be successful the San Jose scale was a most dreaded nursery and orchard foe. It was even thought necessary to destroy infected trees. The lime-sulphur mixture and some other sulphur washes not only kill the San Jose scale but are also useful in reducing fungous injury.

There are several ways of making the lime-sulphur mixture. It is generally best to buy a prepared mixture from some trustworthy dealer. If you find the scale on your trees, write to your state experiment station for directions for combating it.

=The Codling Moth.= The codling moth attacks the apple and often causes a loss of from twenty-five to seventy-five per cent of the crop. In the state of New York this insect is causing an annual loss of about three million dollars. The effect it has on the fruit is most clearly seen in Fig. 152. The moth lays its egg upon the young leaves just after the falling of the blossom. She flies on from apple to apple, depositing an egg each time until from fifty to seventy-five eggs are deposited. The larva, or "worm," soon hatches and eats its way into the apple. Many affected apples ripen too soon and drop as "windfalls." Others remain on the tree and become the common wormy apples so familiar to growers. The larva that emerges from the windfalls moves generally to a tree, crawls up the trunk, and spins its cocoon under a ridge in the bark. From the cocoon the moth comes ready to start a new generation. The last generation of the larv?spends the winter in the cocoon.

Treatment. Destroy orchard trash which may serve as a winter home. Scrape all loose bark from the tree. Spray the tree with arsenate of lead as soon as the flowers fall. A former method of fighting this pest was as follows: bands of burlap four inches wide tied around the tree furnished a hiding-place for larv?that came from windfalls or crawled from wormy apples on the tree. The larv?caught under the bands were killed every five or six days. We know now, however, that a thorough spraying just after the blossoms fall kills the worms and renders the bands unnecessary. Furthermore, spraying prevents wormy apples, while banding does not. Follow the first spraying by a second two weeks later.

It is best to use lime-sulphur mixture or the Bordeaux mixture with arsenate of lead for a spray. Thus one spraying serves against both fungi and insects.

=The Plum Curculio.= The plum curculio, sometimes called the plum weevil, is a little creature about one fifth of an inch long. In spite of its small size the curculio does, if neglected, great damage to our fruit crop. It injures peaches, plums, and cherries by stinging the fruit as soon as it is formed. The word "stinging" when applied to insects--- and this case is no exception--means piercing the object with the egg-layer (ovipositor) and depositing the egg. Some insects occasionally use the ovipositor merely for defense. The curculio has an especially interesting method of laying her egg. First she digs a hole, in which she places the egg and pushes it well down. Then with her snout she makes a crescent-shaped cut in the skin of the plum, around the egg. This mark is shown in Fig. 154. As this peculiar cut is followed by a flow of gum, you will always be able to recognize the work of the curculio. Having finished with one plum, this industrious worker makes her way to other plums until her eggs are all laid. The maggotlike larva soon hatches, burrows through the fruit, and causes it to drop before ripening. The larva then enters the ground to a depth of several inches. There it becomes a pupa, and later, as a mature beetle, emerges and winters in cracks and crevices.

Treatment. Burn orchard trash which may serve as winter quarters. Spraying with arsenate of lead, using two pounds of the mixture to fifty gallons of water, is the only successful treatment for the curculio. For plums and peaches, spray first when the fruit is free from the calyx caps, or dried flower-buds. Repeat the spraying two weeks later. For late peaches spray a third time two weeks after the second spraying. This poisonous spray will kill the beetles while they are feeding or cutting holes in which to lay their eggs.

Fowls in the orchard do good by capturing the larv?before they can burrow, while hogs will destroy the fallen fruit before the larv?can escape.

=The Grape Phylloxera.= The grape phylloxera is a serious pest. You have no doubt seen its galls upon the grape leaf. These galls are caused by a small louse, the phylloxera. Each gall contains a female, which soon fills the gall with eggs. These hatch into more females, which emerge and form new galls, and so the phylloxera spreads (see Fig. 155).

Treatment. The Clinton grape is most liable to injury from this pest. Hence it is better to grow other more resistant kinds. Sometimes the lice attack the roots of the grape vines. In many sections where irrigation is practiced the grape rows are flooded when the lice are thickest. The water drowns the lice and does no harm to the vines.

=The Cankerworm.= The cankerworm is the larva of a moth. Because of its peculiar mode of crawling, by looping its body, it is often called the looping worm or measuring worm (Fig. 157, c). These worms are such greedy eaters that in a short time they can so cut the leaves of an orchard as to give it a scorched appearance. Such an attack practically destroys the crop and does lasting injury to the tree. The worms are green or brown and are striped lengthwise. If the tree is jarred, the worm has a peculiar habit of dropping toward the ground on a silken thread of its own making (Fig. 156).

In early summer the larv?burrow within the earth and pupate there; later they emerge as adults (Fig. 157, d and e). You observe the peculiar difference between the wingless female, d, and the winged male, e. It is the habit of this wingless female to crawl up the trunk of some near-by tree in order to deposit her eggs upon the twigs. These eggs (shown at a and b) hatch into the greedy larv?that do so much damage to our orchards.

Nearly all the common birds feed freely upon the cankerworm, and benefit the orchard in so doing. The chickadee is perhaps the most useful. A recent writer is very positive that each chickadee will devour on an average thirty female cankerworm moths a day; and that if the average number of eggs laid by each female is one hundred and eighty-five, one chickadee would thus destroy in one day five thousand five hundred and fifty eggs, and, in the twenty-five days in which the cankerworm moths crawl up the tree, would rid the orchard of one hundred and thirty-eight thousand seven hundred and fifty. These birds also eat immense numbers of cankerworm eggs before they hatch into worms.

Treatment. The inability of the female to fly gives us an easy way to prevent the larval offspring from getting to the foliage of our trees, for we know that the only highway open to her or her larv?leads up the trunk. We must obstruct this highway so that no crawling creature may pass. This is readily done by smoothing the bark and fitting close to it a band of paper, and making sure that it is tight enough to prevent anything from crawling underneath. Then smear over the paper something so sticky that any moth or larva that attempts to pass will be entangled. Printer's ink will do very well, or you can buy either dendrolene or tanglefoot.

Encourage the chickadee and all other birds, except the English sparrow, to stay in your orchard. This is easily done by feeding and protecting them in their times of need.

=The Apple-Tree Tent Caterpillar.= The apple-tree tent caterpillar is a larva so well known that you only need to be told how to guard against it. The mother of this caterpillar is a reddish moth. This insect passes the winter in the egg state securely fastened on the twig.s

Treatment. There are three principal methods, (1) Destroy the eggs. The egg masses are readily seen in winter and may easily be collected and burned by boys. The chickadee eats great quantities of these eggs. (2) With torches burn the nests at dusk when all the worms are within. You must be very careful in burning or you will harm the young branches with their tender bark. (3) Encourage the residence of birds. Urge your neighbors to make war on the larv? too, since the pest spreads rapidly from farm to farm. Regularly sprayed orchards are rarely troubled by this pest.

=The Twig Girdler.= The twig girdler lays her eggs in the twigs of pear, pecan, apple, and other trees. It is necessary that the larv?develop in dead wood. This the mother provides by girdling the twig so deeply that it will die and fall to the ground.

Treatment. Since the larv?spend the winter in the dead twigs, burn these twigs in autumn or early spring and thus destroy the pest.

=The Peach-Tree Borer.= In Fig. 161 you see the effect of the peach-tree borer's activity. These borers often girdle and thereby kill a tree. Fig. 162 shows the adult state of the insect. The eggs are laid on peach or plum trees near the ground. As soon as the larva emerges, it bores into the bark and remains there for months, passing through the pupa stage before it comes out to lay eggs for another generation.

Treatment. If there are only a few trees in the orchard, digging the worms out with a knife is the best way of destroying them. You can know of the borer's presence by the exuding gum often seen on the tree-trunk. If you pile earth around the roots early in the spring and remove it in the late fall, the winter freezing and thawing will kill many of the larv?

=EXERCISE=

How many apples per hundred do you find injured by the codling moth? Collect some cocoons from a pear or an apple tree in winter, place in a breeding-cage, and watch for the moths that come out. Do you ever see the woodpecker hunting for these same cocoons? Can you find cocoons that have been emptied by this bird? Estimate how many he considers a day's ration. How many apples does he thus save?

Watch the curculio lay her eggs in the plums, peaches, or cherries. What per cent of fruit is thus injured? Estimate the damage. Let the school offer a prize for the greatest number of tent-caterpillar eggs. Watch such trees as the apple, the wild and the cultivated cherry, the oak, and many others.

Make a collection of insects injurious to orchard fruits, showing in each case the whole life history of the insect, that is, eggs, larva, pupa, and the mature insects.

SECTION XXXIII. GARDEN AND FIELD INSECTS

=The Cabbage Worm.= The cabbage worm of the early spring garden is a familiar object, but you may not know that the innocent-looking little white butterflies hovering about the cabbage patch are laying eggs which are soon to hatch and make the dreaded cabbage worms. In Fig. 164 a and b show the common cabbage butterfly, c shows several examples of the caterpillar, and d shows the pupa case. In the pupa stage the insects pass the winter among the remains of old plants or in near-by fences or in weeds or bushes. Cleaning up and burning all trash will destroy many pup?and thus prevent many cabbage worms. In Fig. 164 e and f show the moth and zebra caterpillar; g represents a moth which is the parent of the small green worm shown at h. This worm is a common foe of the cabbage plant.

Treatment. Birds aid in the destruction of this pest. Paris green mixed with air-slaked lime will also kill many larv? After the cabbage has headed, it is very difficult to destroy the worm, but pyrethrum insect powder used freely is helpful.

=The Chinch Bug.= The chinch bug, attacking as it does such important crops as wheat, corn, and grasses, is a well-known pest. It probably causes more money loss than any other garden or field enemy. In Orange county, North Carolina, farmers were once obliged to suspend wheat-growing for two years on account of the chinch bug. In one year in the state of Illinois this bug caused a loss of four million dollars.

Treatment. Unfortunately we cannot prevent all of the damage done by chinch bugs, but we can diminish it somewhat by good clean agriculture. Destroy the winter homes of the insect by burning dry grass, leaves, and rubbish in fields and fence rows. Although the insect has wings, it seldom or never uses them, usually traveling on foot; therefore a deep furrow around the field to be protected will hinder or stop the progress of an invasion. The bugs fall into the bottom of the furrow, and may there be killed by dragging a log up and down the furrow. Write to the Division of Entomology, Washington, for bulletins on the chinch bug. Other methods of prevention are to be found in these bulletins.

The Plant Louse.= The plant louse is very small, but it multiplies with very great rapidity. During the summer the young are born alive, and it is only toward fall that eggs are laid. The individuals that hatch from eggs are generally wingless females, and their young, born alive, are both winged and wingless. The winged forms fly to other plants and start new colonies. Plant lice mature in from eight to fourteen days.

The plant louse gives off a sweetish fluid of which some ants are very fond. You may often see the ants stroking these lice to induce them to give off a freer flow of the "honey dew." This is really a method of milking. However friendly and useful these "cows" may be to the ant, they are enemies to man in destroying so many of his plants.

Treatment. These are sucking insects. Poisons therefore do not avail. They may be killed by spraying with kerosene emulsion or a strong soap solution or with tobacco water. Lice on cabbages are easily killed by a mixture of one pound of lye soap in four gallons of warm water.

=The Squash Bug.= The squash bug does its greatest damage to young plants. To such its attack is often fatal. On larger plants single leaves may die. This insect is a serious enemy to a crop and is particularly difficult to get rid of, since it belongs to the class of sucking insects, not to the biting insects. For this reason poisons are useless.

Treatment. About the only practicable remedy is to pick these insects by hand. We can, however, protect our young plants by small nettings and thus tide them over the most dangerous period of their lives. These bugs greatly prefer the squash as food. You can therefore diminish their attack on your melons, cucumbers, etc. by planting among the melons an occasional squash plant as a "trap plant." Hand picking will be easier on a few trap plants than over the whole field. A small board or large leaf laid beside the young plant often furnishes night shelter for the bugs. The bugs collected under the board may easily be killed every morning.

=The Flea-Beetle.= The flea-beetle inflicts much damage on the potato, tomato, eggplant, and other garden plants. The accompanying figure shows the common striped flea-beetle which lives on the tomato. The larva of this beetle lives inside of the leaves, mining its way through the leaf in a real tunnel. Any substance disagreeable to the beetle, such as plaster, soot, ashes, or tobacco, will repel its attacks on the garden crops.

=The Weevil.= The weevil is commonly found among seeds. Its attacks are serious, but the insect may easily be destroyed.

Treatment. Put the infected seeds in an air-tight box or bin, placing on the top of the pile a dish containing carbon disulphide, a tablespoonful to a bushel of seeds. The fumes of this substance are heavy and will pass through the mass of seeds below and kill all the weevils and other animals there. The bin should be closely covered with canvas or heavy cloth to prevent the fumes from being carried away by the air. Let the seeds remain thus from two to five days. Repeat the treatment if any weevils are found alive. Fumigate when the temperature is 70?Fahrenheit or above. In cold weather or in a loose bin the treatment is not successful. Caution: Do not approach the bin with a light, since the fumes of the chemical used are highly inflammable.

=The Hessian Fly.= The Hessian fly does more damage to the wheat crop than all other insects combined, and probably ranks next to the chinch bug as the second worst insect enemy of the farmer. It was probably introduced into this country by the Hessian troops in the War of the Revolution.

In autumn the insect lays its eggs in the leaves of the wheat. These hatch into the larv? which move down into the crown of the plant, where they pass the winter. There they cause on the plant a slight gall formation, which injures or kills the plant. In the spring adult flies emerge and lay eggs. The larv?that hatch feed in the lower joints of the growing wheat and prevent its proper growth. These larv?pupate and remain as pup?in the wheat stubble during the summer. The fall brood of flies appears shortly before the first heavy frost.

Treatment. Burn all stubble and trash during July and August. If the fly is very bad, it is well to leave the stubble unusually high to insure a rapid spread of the fire. Burn refuse from the threshing-machine, since this often harbors many larv?or pup? Follow the burning by deep plowing, because the burning cannot reach the insects that are in the base of the plants. Delay the fall planting until time for heavy frosts.

=The Potato Beetle; Tobacco Worm.= The potato beetle, tobacco worm, etc., are too well known to need description. Suffice it to say that no good farmer will neglect to protect his crop from any pest that threatens it.

The increase, owing to various causes, of insects, of fungi, of bacterial diseases, makes a study of these pests, of their origin, and of their prevention a necessary part of a successful farmer's training. Tillage alone will no longer render orchard, vineyard, and garden fruitful. Protection from every form of plant enemies must be added to tillage.

In dealing with plants, as with human beings, the great object should be not the cure but the prevention of disease. If disease can be prevented, it is far too costly to wait for it to develop and then to attempt its cure. Men of science are studying the new forms of diseases and new insects as fast as they appear. These men are finding ways of fighting old and new enemies. Young people who expect to farm should early learn to follow their advice.

=EXERCISE=

How does the squash bug resemble the plant louse? Is this a true bug? Gather some eggs and watch the development of the insects in a breeding-cage. Estimate the damage done to some crops by the flea-beetle. What is the best method of prevention?

Do you know the large moth that is the mother of the tobacco worm? You may often see her visiting the blossoms of the Jimson weed. Some tobacco-growers cultivate a few of these weeds in a tobacco field. In the blossom they place a little cobalt or "fly-stone" and sirup. When the tobacco-worm moth visits this flower and sips the poisoned nectar, she will of course lay no more troublesome eggs.

SECTION XXXIV. THE COTTON-BOLL WEEVIL

So far as known, the cotton-boll weevil, an insect which is a native of the tropics, crossed the Rio Grande River into Texas in 1891 and 1892. It settled in the cotton fields around Brownsville. Since then it has widened its destructive area until now it has invaded the whole territory shown by the map on page 177.

This weevil is a small gray or reddish-brown snout-beetle hardly over a quarter of an inch in length. In proportion to its length it has a long beak. It belongs to a family of beetles which breed in pods, in seeds, and in stalks of plants. It is a greedy eater, but feeds only on the cotton plant.

The grown weevils try to outlive the cold of winter by hiding snugly away under grass clumps, cotton-stalks, rubbish, or under the bark of trees. Sometimes they go down into holes in the ground. A comfortable shelter is often found in the forests near the cotton fields, especially in the moss on the trees. The weevils can stand a good deal of cold, but fortunately many are killed by winter weather. Moreover birds destroy many; hence by spring the last year's crop is very greatly diminished.

In the spring, generally about the time cotton begins to form "squares," the weevils shake off their long winter sleep and enter the cotton fields with appetites as sharp as razors. Then shortly the females begin to lay eggs. At first these eggs are laid only in the squares, and generally only one to the square. The young grub hatches from these eggs in two or three days. The newly hatched grub eats the inside of the square, and the square soon falls to the ground. Entire fields may at times be seen without a single square on the plants. Of course no fruit can be formed without squares.

In from one to two weeks the grub or larva becomes fully grown and, without changing its home, is transformed into the pupa state. Then in about a week more the pup?come out as adult weevils and attack the bolls. They puncture them with their snouts and lay their eggs in the bolls. The young grubs, this time hatching out in the boll, remain there until grown, when they emerge through holes that they make. These holes allow dampness to enter and destroy the bolls. This life-round continues until cold weather drives the insects to their winter quarters. By that time they have increased so rapidly that there is often one for every boll in the field.

This weevil is proving very hard to destroy. At present there seem but few ways to fight it. One is to grow cotton that will mature too early for the weevils to do it much harm. A second is to kill as many weevils as possible by burning the homes that shelter them in winter.

The places best adapted for a winter home for the weevil are trash piles, rubbish, driftwood, rotten wood, weeds, moss on trees, etc. A further help, therefore, in destroying the weevil is to cut down and burn all cotton-stalks as soon as the cotton is harvested.

This destroys countless numbers of larv?and pup?in the bolls and greatly reduces the number of weevils. In addition, all cornstalks, all trash, all large clumps of grass in neighboring fields, should be burned, so as to destroy these winter homes of the weevil. Also avoid planting cotton near trees. The bark, moss, and fallen leaves of the tree furnish a winter shelter for the weevils.

A third help in destroying the weevil is to rotate crops. If cotton does not follow cotton, the weevil has nothing on which to feed the second year.

In adopting the first method mentioned the cotton growers have found that by the careful selection of seed, by early planting, by a free use of fertilizers containing phosphoric acid, and by frequent plowing, they can mature a crop about thirty days earlier than they usually do. In this way a good crop can be harvested before the weevils are ready to be most destructive.

CHAPTER VIII

FARM CROPS

Every crop of the farm has been changed and improved in many ways since its forefathers were wild plants. Those plants that best serve the needs of the farmer and of farm animals have undergone the most changes and have received also the greatest care and attention in their production and improvement.

While we have many different kinds of farm crops, the cultivated soil of the world is occupied by a very few. In our country the crop that is most valuable and that occupies the greatest land area is generally known as the grass crop. Included in the general term "grass crop" are the grasses and clovers that are used for pasturage as well as for hay. Next to grass in value come the great cereal, corn, and the most important fiber crop, cotton, closely followed by the great bread crop, wheat. Oats rank fifth in value, potatoes sixth, and tobacco seventh. (These figures are for 1913.)

Success in growing any crop is largely due to the suitableness of soil and climate to that crop. When the planter selects both the most suitable soil and the most suitable climate for each crop, he gets not only the most bountiful yield from the crop but, in addition, he gets the most desirable quality of product. A little careful observation and study soon teach what kinds of soil produce crops of the highest excellence. This learned, the planter is able to grow in each field the several crops best adapted to that special type of soil. Thus we have tobacco soils, trucking soils, wheat and corn soils. Dairying can be most profitably followed in sections where crops like cowpeas, clover, alfalfa, and corn are peculiarly at home. No one should try to grow a new crop in his section until he has found out whether the crop which he wants to grow is adapted to his soil and his climate.

The figures below give the average amount of money made annually an acre on our chief crops:

Flowers and plants, $1911; nursery products, $261; onions, $140; sugar cane, $55; small fruits, $110; hops, $175; vegetables, $78; tobacco, $80; sweet potatoes, $55; hemp, $53; potatoes, $78; sugar beets, $54; sorghum cane, $22; cotton, $22; orchard fruits, $110; peanuts, $21; flax-seed, $14; cereals, $14; hay and forage, $11; castor beans, $6 (United States Census Report).

SECTION XXXV. COTTON

Although cotton was cultivated on the Eastern continent before America was discovered, this crop owes its present kingly place in the business world to the zeal and intelligence of its American growers. So great an influence does it wield in modern industrial life that it is often called King Cotton. Thousands upon thousands of people scan the newspapers each day to see what price its staple is bringing. From its bounty a vast army of toilers, who plant its seed, who pick its bolls, who gin its staple, who spin and weave its lint, who grind its seed, who refine its oil, draw daily bread. Does not its proper production deserve the best thought that can be given it?

In the cotton belt almost any well-drained soil will produce cotton. The following kinds of soil are admirably suited to this plant: red and gray loams with good clay subsoil; sandy soils over clay and sandstone and limestone; rich, well-drained bottom-lands. The safest soils are medium loams. Cotton land must always be well drained.

Cotton was originally a tropical plant, but, strange to say, it seems to thrive best in temperate zones. The cotton plant does best, according to Newman, in climates which have (1) six months of freedom from frost; (2) a moderate, well-distributed rainfall during the plant's growing period; and (3) abundant sunshine and little rain during the plant's maturing period.

In America the Southern states from Virginia to Texas have these climatic qualities, and it is in these states that the cotton industry has been developed until it is one of the giant industries of the world. This development has been very rapid. As late as 1736 the cotton plant was grown as an ornamental flowering plant in many front yards; in 1911, 16,250,276 bales of cotton were grown in the South. In recent years the soil and climate of lower California and parts of Arizona and New Mexico have been found well adapted to cotton.

There are a great many varieties of cotton. Two types are mainly grown by the practical American farmer. These are the short-stapled, upland variety most commonly grown in all the Southern states, and the beautiful, long-stapled, black-seeded sea-island type that grows upon the islands and a portion of the mainland of Georgia, South Carolina, and Florida. The air of the coast seems necessary for the production of this latter variety. The seeds of the sea-island cotton are small, smooth, and black. They are so smooth and stick so loosely to the lint that they are separated from it by roller-gins instead of by saw-gins. When these seeds are planted away from the soil and air of their ocean home, the plant does not thrive.

Many attempts have been made and are still being made to increase the length of the staple of the upland types. The methods used are as follows: selection of seed having a long fiber; special cultivation and fertilization; crossing the short-stapled cotton on the long-stapled cotton. This last process, as already explained, is called hybridizing. Many of these attempts have succeeded, and there are now a large number of varieties which excel the older varieties in profitable yield. The new varieties are each year being more widely grown. Every farmer should study the new types and select the one that will best suit his land. The new types have been developed under the best tillage. Therefore if a farmer would keep the new type as good as it was when he began to grow it, he must give it the same good tillage, and practice seed-selection.

The cotton plant is nourished by a tap-root that will seek food as deeply as loose earth will permit the root to penetrate; hence, in preparing land for this crop the first plowing should be done at least with a two-horse plow and should be deep and thorough. This deep plowing not only allows the tap-root to penetrate, but it also admits a circulation of air.

On some cotton farms it is the practice to break the land in winter or early spring and then let it lie naked until planting-time. This is not a good practice. The winter rains wash more plant food out of unprotected soil than a single crop would use. It would be better, in the late summer or fall, to plant crimson clover or some other protective and enriching crop on land that is to be planted in cotton in the spring. This crop, in addition to keeping the land from being injuriously washed, would greatly help the coming cotton crop by leaving the soil full of vegetable matter.

In preparing for cotton-planting, first disk the land thoroughly, then break with a heavy plow and harrow until a fine and mellow seed-bed is formed. Do not spare the harrow at this time. It destroys many a weed that, if allowed to grow, would have to be cut by costly hoeing. Thorough work before planting saves much expensive work in the later days of the crop. Moreover, no man can afford to allow his plant food and moisture to go to nourish weeds, even for a short time.

The rows should be from three to four feet apart. The width depends upon the richness of the soil. On rich land the rows should be at least four feet apart. This width allows the luxuriant plant to branch and fruit well. On poorer lands the distance of the rows should not be so great. The distribution of the seed in the row is of course most cheaply done by the planter. As a rule it is best not to ridge the land for the seed. Flat culture saves moisture and often prevents damage to the roots. In some sections, however, where the land is flat and full of water, ridging seems necessary if the land cannot be drained.

The cheapest way of cultivating a crop is to prevent grass and weeds from rooting, not to wait to destroy them after they are well rooted. To do this, it is well to run the two-horse smoothing-harrow over the land, across the rows, a few days after the young plants are up. Repeat the harrowing in six or eight days. In addition to destroying the young grass and weeds, this harrowing also removes many of the young cotton plants and thereby saves much hoeing at "chopping-out" time. When the plants are about two inches high they are "chopped out" to secure an evenly distributed stand. It has been the custom to leave two stalks to a hill, but many growers are now leaving only one.

The number of times the crop has to be worked depends on the soil and the season. If the soil is dry and porous, cultivate as often as possible, especially after each rain. Never allow a crust to form after a rain; the roots of plants must have air. Cultivation after each rain forms a dry mulch on the top of the soil and thus prevents rapid evaporation of moisture.

If the fiber (the lint) only is removed from the land on which cotton is grown, cotton is the least exhaustive of the great crops grown in the United States. According to some recent experiments an average crop of cotton removes in the lint only 2.75 pounds of nitrogen, phosphoric acid, potash, lime, and magnesia per acre, while a crop of ten bushels of wheat per acre removes 32.36 pounds of the same elements of plant food. Inasmuch as this crop takes so little plant food from the soil, the cotton-farmer has no excuse for allowing his land to decrease in productiveness. Two things will keep his land in bounteous harvest condition: first, let him return the seeds in some form to the land, or, what is better, feed the ground seeds to cattle, make a profit from the cattle, and return manure to the land in place of the seeds; second, at the last working, let him sow some crop like crimson clover or rye in the cotton rows to protect the soil during the winter and to leave humus in the ground for the spring.

The stable manure, if that is used, should be broadcasted over the fields at the rate of six to ten tons an acre. If commercial fertilizers are used, it may be best to make two applications. To give the young plants a good start, apply a portion of the fertilizer in the drill just before planting. Then when the first blooms appear, put the remainder of the fertilizer in drills near the plants but not too close. Many good cotton-growers, however, apply all the fertilizer at one time.

Relation of Stock to the Cotton Crop. On many farms much of the money for which the cotton is sold in the fall has to go to pay for the commercial fertilizer used in growing the crop. Should not this fact suggest efforts to raise just as good crops without having to buy so much fertilizer? Is there any way by which this can be done? The following suggestions may be helpful. Raise enough stock to use all the cotton seed grown on the farm. To go with the food made from the cotton seed, grow on the farm pea-vine hay, clover, alfalfa, and other such nitrogen-gathering crops. This can be done at small cost. What will be the result?

First, to say nothing of the money made from the cattle, the large quantity of stable manure saved will largely reduce the amount of commercial fertilizer needed. The cotton-farmer cannot afford to neglect cattle-raising. The cattle sections of the country are likely to make the greatest progress in agriculture, because they have manure always on hand.

Second, the nitrogen-gathering crops, while helping to feed the stock, also reduce the fertilizer bills by supplying one of the costly elements of the fertilizer. The ordinary cotton fertilizer consists principally of nitrogen, of potash, and of phosphoric acid. Of these three, by far the most costly is nitrogen. Now peas, beans, clover, and peanuts will leave enough nitrogen in the soil for cotton, so that if they are raised, it is necessary to buy only phosphoric acid and sometimes potash.

SECTION XXXVI. TOBACCO

The tobacco plant connects Indian agriculture with our own. It has always been a source of great profit to our people. In the early colonial days tobacco was almost the only money crop. Many rich men came to America in those days merely to raise tobacco.

Although tobacco will grow in almost any climate, the leaves, which, as most of you know, are the salable part of the plant, get their desirable or undesirable qualities very largely from the soil and from the climate in which they grow.

The soil in which tobacco thrives best is one which has the following qualities: dryness, warmth, richness, depth, and sandiness.

Commercial fertilizers also are almost a necessity; for, as tobacco land is limited in area, the same land must be often planted in tobacco. Hence even a fresh, rich soil that did not at first require fertilizing soon becomes exhausted, and, after the land has been robbed of its plant food by crop after crop of tobacco, frequent application of fertilizers and other manures becomes necessary. However, even tobacco growers should rotate their crops as much as possible.

Deep plowing--from nine to thirteen inches--is also a necessity in preparing the land, for tobacco roots go deep into the soil. After this deep plowing, harrow until the soil is thoroughly pulverized and is as fine and mellow as that of the flower-garden.

Unlike most other farm crops the tobacco plant must be started first in a seed-bed. To prepare a tobacco bed the almost universal custom has been to proceed as follows. Carefully select a protected spot. Over this spot pile brushwood and then burn it. The soil will be left dry, and all the weed seeds will be killed. The bed is then carefully raked and smoothed and planted. Some farmers are now preparing their beds without burning. A tablespoonful of seed will sow a patch twenty-five feet square. A cheap cloth cover is put over the bed. If the seeds come up well, a patch of this size ought to furnish transplants for five or six acres. In sowing, it is not wise to cover the seed deeply. A light raking in or an even rolling of the ground is all that is needed.

The time required for sprouting is from two to three weeks. The plants ought to be ready for transplanting in from four to six weeks. Weeds and grass should of course be kept out of the seed-bed.

The plants, when ready, are transplanted in very much the same way as cabbages and tomatoes. The transplanting was formerly done by hand, but an effective machine is now widely used. The rows should be from three to three and a half feet apart, and the plants in the rows about two or three feet apart. If the plants are set so that the plow and cultivator can be run with the rows and also across the rows, they can be more economically worked. Tobacco, like corn, requires shallow cultivation. Of course the plants should be worked often enough to give clean culture and to provide a soil mulch for saving moisture.

In tobacco culture it is necessary to pinch off the "buttons" and to cut off the tops of the main stalk, else much nourishment that should go to the leaves will be given to the seeds. The suckers must also be cut off for the same reason.

The proper time for harvesting is not easily fixed; one becomes skillful in this work only through experience in the field. Briefly, we may say that tobacco is ready to be cut when the leaves on being held up to the sun show a light or golden color, when they are sticky to the touch, and when they break easily on being bent. Plants that are overripe are inferior to those that are cut early.

The operations included in cutting, housing, drying, shipping, sweating, and packing require skill and practice.

SECTION XXXVII. WHEAT

Wheat has been cultivated from earliest times. It was a chief crop in Egypt and Palestine, and still holds its importance in the temperate portions of Europe, Asia, Africa, Australia, and America.

This crop ranks third in value in the United States. It grows in cool, in temperate, and in warm climates, and in many kinds of soil. It does best in clay loam, and worst in sandy soils. Clogged and water-soaked land will not grow wheat with profit to the farmer; for this reason, where good wheat-production is desired the soil must be well drained and in good physical condition--that is, the soil must be open, crumbly, and mellow.

Clay soils that are hard and lifeless can be made valuable for wheat-production by covering the surface with manure, by good tillage, and by a thorough system of crop-rotation. Cowpeas and other legumes make a most valuable crop to precede wheat, for in growing they add atmospheric nitrogen to the soil, and their roots loosen the root-bed, thereby admitting a free circulation of air and adding humus to the soil. Moreover, the legumes leave the soil with its grains fairly close packed, and this is a help in wheat growing.

One may secure a good seed-bed after cotton and corn as well as after cowpeas and other legumes. They are summer-cultivated crops, and the clean culture that has been given them renders the surface soil mellow and the undersoil firm and compact. They are not so good, however, as cowpeas, since they add no atmospheric nitrogen to the soil, as all leguminous crops do.

From one to two inches is the most satisfactory depth for planting wheat. The largest number of seeds comes up when planted at this depth. A mellow soil is very helpful to good coming up and provides a most comfortable home for the roots of the plant. A compact soil below makes a moist undersoil; and this is desirable, for the soil water is needed to dissolve plant food and to carry it up through the plant, where it is used in building tissue.

There are a great many varieties of wheat: some are bearded, others are smooth; some are winter and others are spring varieties. The smooth-headed varieties are most agreeable to handle during harvest and at threshing-time. Some of the bearded varieties, however, do so well in some soils and climates that it is desirable to continue growing them, though they are less agreeable to handle. No matter what variety you are accustomed to raise, it may be improved by careful seed-selection.

The seed-drill is the best implement for planting wheat. It distributes the grains evenly over the whole field and leaves the mellow soil in a condition to catch what snow may fall and secure what protection it affords.

In many parts of the country, because not enough live stock is raised, there is often too little manure to apply to the wheat land. Where this is the case commercial fertilizers must be used. Since soils differ greatly, it is impossible to suggest a fertilizer adapted to all soils. The elements usually lacking in wheat soils are nitrogen, phosphoric acid, and potash. The land may be lacking in one of these plant foods or in all; in either case a maximum crop cannot possibly be raised. The section on manuring the soil will be helpful to the wheat-grower.

It should be remembered always in buying fertilizers for wheat that whenever wheat follows cowpeas or clover or other legumes there is seldom need of using nitrogen in the fertilizer; the tubercles on the pea or clover roots will furnish that. Hence, as a rule, only potash and phosphoric acid will have to be purchased as plant food.

The farmer is assisted always by a study of his crop and by a knowledge of how it grows. If he find the straw inferior and short, it means that the soil is deficient in nitrogen; but on the other hand, if the straw be luxuriant and the heads small and poorly filled, he may be sure that his soil contains too little phosphoric acid and potash.

=EXERCISE=

Let the pupils secure several heads of wheat and thresh each separately by hand. The grains should then be counted and their plumpness and size observed. The practical importance of this is obvious, for the larger the heads and the greater the number of grains, the larger the yield per acre. Let them plant some of the large and some of the small grains. A single test of this kind will show the importance of careful seed-selection.

SECTION XXXVIII. CORN

When the white man came to this country he found the Indians using corn; for this reason, in addition to its name maize, it is called Indian corn. Before that time the civilized world did not know that there was such a crop. The increase in the yield and the extension of the acres planted in this strictly American crop have kept pace with the rapid and wonderful growth of our country. Corn is king of the cereals and the most important crop of American agriculture. It grows in almost every section of America. There is hardly any limit to the uses to which its grain and its stalks are now put. Animals of many kinds are fed on rations into which it enters. Its grains in some form furnish food to more people than does any other crop except possibly rice. Its stalk and its cob are manufactured into many different and useful articles.

A soil rich in either decaying animal or vegetable matter, loose, warm, and moist but not wet, will produce a better crop of corn than any other. Corn soil should always be well tilled and cultivated.

The proper time to begin the cultivation of corn is before it is planted. Plow well. A shallow, worn-out soil should not be used for corn, but for cowpeas or rye. After thorough plowing, the harrow--either the disk or spring-tooth-- should be used to destroy all clods and leave the surface mellow and fine. The best results will be obtained by turning under a clover sod that has been manured from the savings of the barnyard.

When manure is not available, commercial fertilizers will often prove profitable on poor lands. Careful trials will best determine how much fertilizer to an acre is necessary, and what kinds are to be used. A little study and experimenting on the farmer's part will soon enable him to find out both the kind and the amount of fertilizer that is best suited to his land.

The seed for this crop should be selected according to the plan suggested in Section XIX.

The most economical method of planting is by means of the horse planter, which, according to its adjustment, plants regularly in hills or in drills. A few days after planting, the cornfield should be harrowed with a fine-tooth harrow to loosen the top soil and to kill the grass and the weed seeds that are germinating at the surface. When the corn plants are from a half inch to an inch high, the harrow may again be used. A little work before the weeds sprout will save many days of labor during the rest of the season, and increase the yield.

Corn is a crop that needs constant cultivation, and during the growing season the soil should be stirred at least four times. This cultivation is for three reasons:

1. To destroy weeds that would take plant food and water.

2. To provide a mulch of dry soil so as to prevent the evaporation of moisture. The action of this mulch has already been explained.

3. Because "tillage is manure." Constant stirring of the soil allows the air to circulate in it, provides a more effective mulch, and helps to change unavailable plant food into the form that plants use.

Deep culture of corn is not advisable. The roots in their early stages of growth are shallow feeders and spread widely only a few inches below the surface. The cultivation that destroys or disturbs the roots injures the plants and lessens the yield. We cultivate because of the three reasons given above, and not to stir the soil about the roots or to loosen it there.

In many parts of the country the cornstalks are left standing in the fields or are burned. This is a great mistake, for the stalks are worth a good deal for feeding horses, cattle, and sheep. These stalks may always be saved by the use of the husker and shredder. Corn after being matured and cut can be put in shocks and left thus until dry enough to run through the husker and shredder.

This machine separates the corn from the stalk and husks it. At the same time it shreds tops, leaves, and butts into a food that is both nutritious and palatable to stock. For the amount that animals will eat, almost as much feeding value is obtained from corn stover treated in this way as from timothy hay. The practice of not using the stalks is wasteful and is fast being abandoned. The only reason that so much good food is being left to decay in the field is because so many people have not fully learned the feeding value of the stover.

=EXERCISE=

To show the effect of cultivation on the yield of corn, let the pupils lay off five plats in some convenient field. Each plat need consist of only two rows about twenty feet long. Treat each plat as follows:

Plat 1. No cultivation: let weeds grow.

Plat 2. Mulch with straw.

Plat 3. Shallow cultivation: not deeper than two inches and at least five times during the growing season.

Plat 4. Deep cultivation: at least four inches deep, so as to injure and tear out some of the roots (this is a common method).

Plat 5. Root-pruning: ten inches from the stalk and six inches deep, prune the roots with a long knife. Cultivate five times during the season.

Observe plats during the summer, and at husking-time note results.

SECTION XXXIX. PEANUTS

This plant is rich in names, being known locally as "ground pea," "goober," "earthnut," and "pindar," as well as generally by the name of "peanut." The peanut is a true legume, and, like other legumes, bears nitrogen-gathering tubercles upon its roots. The fruit is not a real nut but rather a kind of pea or bean, and develops from the blossom. After the fall of the blossom the "spike," or flower-stalk, pushes its way into the ground, where the nut develops. If unable to penetrate the soil the nut dies.

In the United States, North and South Carolina, Virginia, and Tennessee have the most favorable climates for peanut culture. Suitable climate and soil, however, may be found from New Jersey to the Mississippi valley. A high, porous, sandy loam is the most suitable. Stiffer soils, which may in some cases yield larger crops than the loams, are yet not so profitable, for stiff soils injure the color of the nut. Lime is a necessity and must be supplied if the soil is deficient. Phosphoric acid and potash are needed.

Greater care than is usually bestowed should be given to the selection of the peanut seed. In addition to following the principles given in Section XVIII, all musty, defective seeds must be avoided and all frosted kernels must be rejected. Before it dries, the peanut seed is easily injured by frost. The slightest frost on the vines, either before or after the plants are dug, does much harm to the tender seed.

In growing peanuts, thorough preparation of the soil is much better than later cultivation. Destroy the crop of young weeds, but do not disturb the peanut crop by late cultivation. Harvest before frost, and shock high to keep the vines from the ground.

The average yield of peanuts in the United States is twenty-two bushels an acre. In Tennessee the yield is twenty-nine bushels an acre, and in North Carolina and Virginia it reaches thirty bushels an acre.

SECTION XL. SWEET POTATOES

The roots of sweet potatoes are put on the market in various forms. Aside from the form in which they are ordinarily sold, some potatoes are dried and then ground into flour, some are canned, some are used to make starch, some furnish a kind of sugar called glucose, and some are even used to make alcohol.

The fact that there are over eighty varieties of potatoes shows the popularity of the plant. Now it is evident that all of these varieties cannot be equally desirable. Hence the wise grower will select his varieties with prudent forethought. He should study his market, his soil, and his seed (see Section XVIII).

Four months of mild weather, months free from frost and cold winds, are necessary for the growing of sweet potatoes. In a mild climate almost any loose, well-drained soil will produce them. A light, sandy loam, however, gives a cleaner potato and one, therefore, that sells better.

The sweet potato draws potash, nitrogen, and phosphoric acid from the soil, but in applying these as fertilizers the grower must study and know his own soil. If he does not he may waste both money and plant food by the addition of elements already present in sufficient quantity in the soil. The only way to come to reliable conclusions as to the needs of the soil is to try two or three different kinds of fertilizers on plats of the same soil, during the same season, and notice the resulting crop of potatoes.

Sweet potatoes will do well after almost any of the usual field crops. This caution, however, should be borne in mind. Potatoes should not follow a sod. This is because sods are often thick with cutworms, one of the serious enemies of the potato.

It is needless to say that the ground must be kept clean by thorough cultivation until the vines take full possession of the field.

In harvesting, extreme care should be used to avoid cutting and bruising the potato, since bruises are as dangerous to a sweet potato as to an apple, and render decay almost a certainty. Lay aside all bruised potatoes for immediate use.

For shipment the potatoes should be graded and packed with care. An extra outlay of fifty cents a barrel often brings a return of a dollar a barrel in the market. One fact often neglected by Southern growers who raise potatoes for a Northern market is that the Northern markets demand a potato that will cook dry and mealy, and that they will not accept the juicy, sugary potato so popular in the South.

The storage of sweet potatoes presents difficulties owing to their great tendency to decay under the influence of the ever-present fungi and bacteria. This tendency can be met by preventing bruises and by keeping the bin free from rotting potatoes. The potatoes should be cleaned, and after the moisture has been dried off they should be stored in a dry, warm place.

The sweet-potato vine makes a fair quality of hay and with proper precaution may be used for ensilage. Small, defective, unsalable potatoes are rich in sugar and starch and are therefore good stock food. Since they contain so much water they must be used only as an aid to other diet.

SECTION XLI. WHITE, OR IRISH, POTATOES

Maize, or Indian corn, and potatoes are the two greatest gifts in the way of food that America has bestowed on the other nations. Since their adoption in the sixteenth century as a new food from recently discovered America, white potatoes have become one of the world's most important crops.

No grower will harvest large crops of potatoes unless he chooses soil that suits the plant, selects his seed carefully, cultivates thoroughly, feeds his land sufficiently, and sprays regularly.

The soil should be free from potato scab. This disease remains in land for several years. Hence if land is known to have any form of scab in it, do not plant potatoes in such land. Select for this crop a deep and moderately light, sandy loam which has an open subsoil and which is rich in humus. The soil must be light enough for the potatoes, or tubers, to enlarge easily and dry enough to prevent rot or blight or other diseases. Potato soil should be so close-grained that it will hold moisture during a dry spell and yet so well-drained that the tubers will not be hurt by too much moisture in wet weather.

If the land selected for potatoes is lacking in humus, fine compost or well-rotted manure will greatly increase the yield. However, it should be remembered that green manure makes a good home for the growth of scab germs. Hence it is safest to apply this sort of manure in the fall, or, better still, use a heavy dressing of manure on the crop which the potatoes are to follow. Leguminous crops supply both humus and nitrogen and, at the same time, improve the subsoil. Therefore such crops are excellent to go immediately before potatoes. If land is well supplied with humus, commercial fertilizers are perhaps safer than manure, for when these fertilizers are used the amount of plant food is more easily regulated. Select a fertilizer that is rich in potash. For gardens unleached wood ashes make a valuable fertilizer because they supply potash. Early potatoes need more fertilization than do late ones. While potatoes do best on rich land, they should not be overfed, for a too heavy growth of foliage is likely to cause blight.

Be careful to select seed from sound potatoes which are entirely free from scab. Get the kinds that thrive best in the section in which they are to be planted and which suit best the markets in which they are to be sold. Seed potatoes should be kept in a cool place so that they will not sprout before planting-time. As a rule consumers prefer a smooth, regularly shaped, shallow-eyed white or flesh-colored potato which is mealy when cooked. Therefore, select seed tubers with these qualities. It seems proved that when whole potatoes are used for seed the yield is larger than when sliced potatoes are planted. It is of course too costly to plant whole potatoes, but it is a good practice to cause the plants to thrive by planting large seed pieces.

Like other crops, potatoes need a thoroughly prepared seed-bed and intelligent cultivation. Break the land deep. Then go over it with an ordinary harrow until all clods are broken and the soil is fine and well closed. The rows should be at least three feet from one another and the seeds placed from twelve to eighteen inches apart in the row, and covered to a depth of three or four inches. A late crop should be planted deeper than an early one. Before the plants come up it is well to go over the field once or twice with a harrow so as to kill all weeds. Do not fail to save moisture by frequent cultivation. After the plants start to grow, all cultivation should be shallow, for the roots feed near the surface and should not be broken. Cultivate as often as needed to keep down weeds and grass and to keep the ground fine.

Allow potatoes to dry thoroughly before they are stored, but never allow them to remain long in the sunshine. Never dig them in damp weather, for the moisture clinging to them will cause them to rot. After the tubers are dry, store them in barrels or bins in a dry, cool, and dark place. Never allow them to freeze.

Among the common diseases and insect pests that attack the leaves and stems of potato vines are early blight, late blight, brown rot, the flea-beetle, and the potato beetle, or potato bug. Spraying with Bordeaux mixture to which a small portion of Paris green has been added will control both the diseases and the pests. The spraying should begin when the plants are five or six inches high and should not cease until the foliage begins to die.

Scab is a disease of the tubers. It may be prevented (1) by using seed potatoes that are free from scab; (2) by planting land in which there is no scab; and (3) by soaking the seed in formalin (see page 135).

SECTION XLII. OATS

The oat plant belongs to the grass family. It is a hardy plant and, under good conditions, a vigorous grower. It stands cold and wet better than any other cereal except possibly rye. Oats like a cool, moist climate. In warm climates, oats do best when they are sowed in the fall. In cooler sections, spring seeding is more generally practiced.

There are a great many varieties of oats. No one variety is best adapted to all sections, but many varieties make fine crops in many sections. Any variety is desirable which has these qualities: power to resist disease and insect enemies, heavy grains, thin hulls, good color, and suitability to local surroundings.

As oats and rye make a better yield on poor land than any other cereals, some farmers usually plant these crops on their poorest lands. However, no land is too good to be used for so valuable a crop as oats. Oats require a great deal of moisture; hence light, sandy soils are not so well adapted to this crop as are the sandy loams and fine clay loams with their closer and heavier texture.

If oats are to be planted in the spring, the ground should be broken in the fall, winter, or early spring so that no delay may occur at seeding-time. But to have a thoroughly settled, compact seed-bed the breaking of the land should be done at least a month before the seeding, and it will help greatly to run over the land with a disk harrow immediately after the breaking.

Oats may be planted by scattering them broadcast or by means of a drill. The drill is better, because the grains are more uniformly distributed and the depth of planting is better regulated. The seeds should be covered from one and a half to two inches deep. In a very dry season three inches may not be too deep. The amount of seed needed to the acre varies considerably, but generally the seeding is from two to three bushels an acre. On poor lands two bushels will be a fair average seeding; on good lands as much as three bushels should be used.

This crop fits in well, over wide areas, with various rotations. As the purpose of all rotation is to keep the soil productive, oats should alternate every few years with one of the nitrogen-gathering crops. In the South, cowpeas, soy beans, clovers, and vetches may be used in this rotation. In the North and West the clovers mixed with timothy hay make a useful combination for this purpose.

Spring-sowed oats, since they have a short growing season, need their nitrogenous plant food in a form which can be quickly used. To supply this nitrogen a top-dressing of nitrate of soda or sulphate of lime is helpful. The plant can gather its food quickly from either of these two. As fall-sowed oats have of course a longer growing season, the nitrogen can be supplied by well-rotted manure, blood, tankage, or fish-scrap. Use barnyard manure carefully. Do not apply too much just before seeding, and use only thoroughly rotted manure. It is always desirable to have a bountiful supply of humus in land on which oats are to be planted.

The time of harvesting will vary with the use which is to be made of the oats. If the crop is to be threshed, the harvesting should be done when the kernels have passed out of the milk into the hard dough state. The lower leaves of the stalks will at this time have turned yellow, and the kernels will be plump and full. Do not, however, wait too long, for if you do the grain will shatter and the straw lose in feeding value.

On the other hand, if the oats are to be cut for hay it is best to cut them while the grains are still in the milk stage. At this stage the leaves are still green and the plants are rich in protein.

Oats should be cured quickly. It is very important that threshed oats should be dry before they are stored. Should they on being stored still contain moisture, they will be likely to heat and to discolor. Any discoloring will reduce their value. Nor should oats ever be allowed to remain long in the fields, no matter how well they may seem to be shocked. The dew and the rain will injure their value by discoloring them more or less.

Oats are muscle-builders rather than fat-formers. Hence they are a valuable ration for work animals, dairy cows, and breeding-stock.

SECTION XLIII. RYE

Rye has the power of gathering its food from a wider area than most other plants. Of course, then, it is a fine crop for poor land, and farmers often plant it only on worn land. However, it is too good a cereal to be treated in so ungenerous a fashion. As a cover-crop for poor land it adds much humus to the soil and makes capital grazing.

There are two types of rye--the winter and the spring. The winter type is chiefly grown in this country. Rye seeds should be bought as near home as possible, for this plant thrives best when the new crop grows under the same conditions as the seed crop.

Rye will grow on almost any soil that is drained. Soils that are too sandy for wheat will generally yield good crops of rye. Clay soils, however, are not adapted to the plant nor to the grazing for which the plant is generally sowed. For winter rye the land should be broken from four to six inches. Harrows should follow the plows until the land is well pulverized. In some cold prairie lands, however, rye is put in with a grain-drill before a plow removes the stubble from the land. The purpose of planting in this way is to let the stubble protect the young plants from cold, driving winds.

Rye should go into the ground earlier than wheat. In cold, bleak climates, as well as on poor land, the seeding should be early. The young plant needs to get rooted and topped before cold weather sets in. The only danger in very early planting is that leaf-rust sometimes attacks the forward crop. Of course the earlier the rye is ready for fall and winter pasturage, the better. If a drill is used for planting, a seeding of from three to four pecks to the acre should give a good stand. In case the seeds are to be sowed broadcast, a bushel or a bushel and a half for every acre is needed. The seed should be covered as wheat seed is and the ground rolled.

Rye is generally used as a grazing or as a soiling crop. Therefore its value will depend largely on its vigorous growth in stems and leaves. To get this growth, liberal amounts of nitrogenous fertilizer will have to be applied unless the land is very rich. Put barnyard manure on the land just after the first breaking and disk the manure into the soil. Acid phosphate and kainite added to the manure may pay handsomely. A spring top-dressing of nitrate of soda is usually helpful.

Rye has a stiff straw and does not fall, or "lodge," so badly as some of the other cereals. As soon as rye that is meant for threshing is cut, it should be put up in shocks until it is thoroughly dry. Begin the cutting when the kernels are in a tough dough state. The grain should never stand long in the shocks.

SECTION XLIV. BARLEY

Barley is one of the oldest crops known to man. The old historian Pliny says that barley was the first food of mankind. Modern man however prefers wheat and corn and potatoes to barley, and as a food this ancient crop is in America turned over to the lower animals. Brewers use barley extensively in making malt liquors. Barley grows in nearly all sections of our country, but a few states--namely, Minnesota, California, Wisconsin, Iowa, and North and South Dakota--are seeding large areas to this crop.

For malting purposes the barley raised on rather light, friable, porous soil is best. Soils of this kind are likely to produce a medium yield of bright grain. Fertile loamy and clay soils make generally a heavier yield of barley, but the grain is dark and fit only to be fed to stock. Barley is a shallow feeder, and can reach only such plant food as is found in the top soil, so its food should always be put within reach by a thorough breaking, harrowing, and mellowing of the soil, and by fertilizing if the soil is poor. Barley has been successfully raised both by irrigation and by dry-farming methods. It requires a better-prepared soil than the other grain crops; it makes fine yields when it follows some crop that has received a heavy dressing of manure. Capital yields are produced after alfalfa or after root crops. This crop usually matures within a hundred days from its seeding.

When the crop is to be sold to the brewers, a grain rich in starch should be secured. Barley intended for malting should be fertilized to this end. Many experiments have shown that a fertilizer which contains much potash will produce starchy barley. If the barley be intended for stock, you should breed so as to get protein in the grain and in the stalk. Hence barley which is to be fed should be fertilized with mixtures containing nitrogen and phosphoric acid. Young barley plants are more likely to be hurt by cold than either wheat or oats. Hence barley ought not to be seeded until all danger from frost is over. The seeds should be covered deeper than the seeds of wheat or of oats. Four inches is perhaps an average depth for covering. But the covering will vary with the time of planting, with the kind of ground, with the climate, and with the nature of the season. Fewer seeds will be needed if the barley is planted by means of a drill.

Like other cereals, barley should not be grown continuously on the same land. It should take its place in a well-planned rotation. It may profitably follow potatoes or other hoed crops, but it should not come first after wheat, oats, or rye.

Barley should be harvested as soon as most of its kernels have reached the hard dough state. It is more likely to shatter its grain than are other cereals, and it should therefore be handled with care. It must also be watched to prevent its sprouting in the shocks. Be sure to put few bundles in the shock and to cap the shock securely enough to keep out dew and rain. If possible the barley should be threshed directly from the shock, as much handling will occasion a serious loss from shattering.

SECTION XLV. SUGAR PLANTS

In the United States there are three sources from which sugar is obtained; namely, the sugar-maple, the sugar-beet, and the sugar-cane. In the early days of our country considerable quantities of maple sirup and maple sugar were made. This was the first source of sugar. Then sugar-cane began to be grown. Later the sugar-beet was introduced.

=Maple Products.= In many states sirup and sugar are still made from maple sap. In the spring when the sap is flowing freely maple trees are tapped and spouts are inserted. Through these spouts the sap flows into vessels set to catch it. The sap is boiled in evaporating-pans, and made into either sirup or sugar. Four gallons of sap yield about one pound of sugar. A single tree yields from two to six pounds of sugar in a season. The sap cannot be kept long after it is collected. Practice and skill are needed to produce an attractive and palatable grade of sirup or of sugar.

=Sugar-Beets.= The sugar-beet is a comparatively new root crop in America.

The amount of sugar that can be obtained from beets varies from twelve to twenty per cent. The richness in sugar depends somewhat on the variety grown and on the soil and the climate.

So far most of our sugar-beet seeds have been brought over from Europe. Some of our planters are now, however, gaining the skill and the knowledge needed to grow these seeds. It is of course important to grow seeds that will produce beets containing much sugar.

These beets do well in a great variety of soils if the land is rich, well prepared, and well drained, and has a porous subsoil.

Beets cannot grow to a large size in hard land. Hence deep plowing is very necessary for this crop. The soil should be loose enough for the whole body of the beet to remain underground. Some growers prefer spring plowing and some fall plowing, but all agree that the land should not be turned less than eight or ten inches. The subsoil, however, should not be turned up too much at the first deep plowing.

Too much care cannot be taken to make the seed-bed firm and mellow and to have it free from clods. If the soil is dry at planting-time and there is likelihood of high winds, the seed-bed may be rolled with profit. Experienced growers use from ten to twelve pounds of seeds to an acre. It is better to use too many rather than too few seeds, for it is easy to thin out the plants, but rather difficult to transplant them. The seeds are usually drilled in rows about twenty inches apart. Of course, if the soil is rather warm and moist at planting-time, fewer seeds will be needed than when germination is likely to be slow.

A good rotation should always be planned for this beet. A very successful one is as follows: for the first year, corn heavily fertilized with stable manure; for the second year, sugar-beets; for the third year, oats or barley; for the fourth year, clover; then go back again to corn. In addition to keeping the soil fertile, there are two gains from this rotation: first, the clean cultivation of the corn crop just ahead of the beets destroys many of the weed seeds; second, the beets must be protected from too much nitrogen in the soil, for an excess of nitrogen makes a beet too large to be rich in sugar. The manure, heavily applied to the corn, will leave enough nitrogen and other plant food in the soil to make a good crop of beets and avoid any danger of an excess.

When the outside leaves of the beet take on a yellow tinge and drop to the ground, the beets are ripe. The mature beets are richer in sugar than the immature, therefore they should not be harvested too soon. They may remain in the ground without injury for some time after they are ripe. Cold weather does not injure the roots unless it is accompanied by freezing and thawing.

The beets are harvested by sugar-beet pullers or by hand. If the roots are to be gathered by hand they are usually loosened by plowing on each side of them. If the roots are stored they should be put in long, narrow piles and covered with straw and earth to protect them from frost. A ventilator placed at the top of the pile will enable the heat and moisture to escape. If the beets get too warm they will ferment and some of their sugar will be lost.

=Sugar-Cane.= Sugar-cane is grown along the Gulf of Mexico and the South Atlantic coast. In Mississippi, in Alabama, Florida, Georgia, South Carolina, northern Louisiana, and in northern Texas it is generally made into sirup. In southern Louisiana and southern Texas the cane is usually crushed for sugar or for molasses.

The sugar-cane is a huge grass. The stalk, which is round, is from one to two inches in thickness.

The stalks vary in color. Some are white, some yellow, some green, some red, some purple, and some black, while others are a mixture of two or three of these colors. As shown in Fig. 214 the stalk has joints at distances of from two to six inches. These joints are called nodes, and the sections between the nodes are known as internodes. The internodes ripen from the roots upward, and as each ripens it casts its leaves. The stalk, when ready for harvesting, has only a few leaves on the top.

Under each leaf and on alternate sides of the cane a bud, or "eye," forms. From this eye the cane is usually propagated; for, while in tropical countries the cane forms seeds, yet these seeds are rarely fertile. When the cane is ripe it is stripped of leaves, topped, and cut at the ground with a knife. The sugar is contained in solution in the pith of the cane.

Cane requires an enormous amount of water for its best growth, and where the rainfall is not great enough, the plants are irrigated. It requires from seventy-five to one hundred gallons of water to make a pound of sugar. Cane does best where there is a rainfall of two inches a week. At the same time a well-drained soil is necessary to make vigorous canes.

The soils suited to this plant are those which contain large amounts of fertilizing material and which can hold much water. In southern Louisiana alluvial loams and loamy clay soils are cultivated. In Georgia, Alabama, and Florida light, sandy soils, when properly fertilized and worked, make good crops.

Cane is usually planted in rows from five to six feet apart. A trench is opened in the center of the row with a plow and in this open furrow is placed a continuous line of stalks which are carefully covered with plow, cultivator, or hoe. From one to three continuous lines of stalks are placed in the furrow. From two to six tons of seed cane are needed for an acre. In favorable weather the cane soon sprouts and cultivation begins. Cane should be cultivated at short intervals until the plants are large enough to shade the soil. In Louisiana one planting of cane usually gives two crops. The first is called plant cane; the second is known as first-year stubble, or ratoon. Sometimes second-year stubble is grown.

In Louisiana large quantities of tankage, cotton-seed meal, and acid phosphate are used to fertilize cane-fields. Each country has its own time for planting and harvesting. In Louisiana, for example, canes are planted from October to April. In the United States cane is harvested each year because of frost, but in tropical countries the stalks are permitted to grow from fifteen to twenty-four months.

On many farms a small mill, the rollers of which are turned by horses, is used for crushing the juice out of the cane. The juice is then evaporated in a kettle or pan. This equipment is very cheap and can easily be operated by a small family. While these mills rarely extract more than one half of the juice in the cane, the sirup made by them is very palatable and usually commands a good price. Costly machinery which saves most of the juice is used in the large commercial sugar houses.

SECTION XLVI. HEMP AND FLAX

In the early ages of the world, mankind is supposed to have worn very little or no clothing. Then leaves and the inner bark of trees were fashioned into a protection from the weather. These flimsy garments were later replaced by skins and furs. As man advanced in knowledge, he learned how to twist wool and hairs into threads and to weave these into durable garments. Still later, perhaps, he discovered that some plants conceal under their outer bark soft, tough fibers that can be changed into excellent cloth. Flax and hemp were doubtless among the first plants to furnish this fiber.

=Flax.= Among the fiber crops of the world, flax ranks next to cotton. It is the material from which is woven the linen for sheets, towels, tablecloths, shirts, collars, dresses, and a host of other articles. Fortunately for man, flax will thrive in many countries and in many climates. The fiber from which these useful articles are made, unlike cotton fiber, does not come from the fruit, but from the stem. It is the soft, silky lining of the bark which lies between the woody outside and the pith cells of the stem.

The Old World engages largely in flax culture and flax manufacture, but in our country flax is grown principally for its seed. From the seeds we make linseed oil, linseed-oil cake, and linseed meal.

Flax grows best on deep, loamy soils, but also makes a profitable growth on clay soils. With sufficient fertilizing material it can be grown on sandy lands. Nitrogen is especially needed by this plant and should be liberally supplied. To meet this demand for nitrogen, it pays to plant a leguminous crop immediately before flax.

After a mellow seed-bed has been made ready and after the weather is fairly warm, sow, if a seed crop is desired, at the rate of from two to three pecks an acre. A good seed crop will not be harvested if the plants are too thick. On the other hand, if a fiber crop is to be raised, it is desirable to plant more thickly, so that the stalks may not branch, but run up into a single stem. From a bushel to two bushels of seed is in this case used to an acre. Flax requires care and work from start to finish.

When the seeds are full and plump the flax is ready for harvesting. In America a binder is generally used for cutting the stalks. Our average yield of flax is from eight to fifteen bushels an acre.

=Hemp.= Like flax, hemp adapts itself wonderfully to many countries and many climates. However, in America most of our hemp is grown in Kentucky.

Hemp needs soil rich enough to give the young plants a very rapid growth in their early days so that they may form long fibers. To give this crop abundant nitrogen without great cost, it should be grown in a rotation which includes one of the legumes. Rich, well-drained bottom-lands produce the largest yields of hemp, but uplands which have been heavily manured make profitable yields.

The ground for hemp is prepared as for other grain crops. The seed is generally broadcasted for a fiber crop and then harrowed in. No cultivation is required after seeding.

If hemp is grown for seed, it is best to plant with a drill so that the crop may be cultivated. The stalks after being cut are put in shocks until they are dry. Then the seeds are threshed. Large amounts of hemp seed are sold for caged birds and for poultry; it is also used for paint-oils.

SECTION XLVII. BUCKWHEAT

Buckwheat shares with rye and cowpeas the power to make a fairly good crop on poor land. At the same time, of course, a full crop can be expected only from fertile land.

The three varieties most grown in America are the common gray, the silver-hull, and the Japanese. The seeds of the common gray are larger than the silver-hull, but not so large as the Japanese. The seeds from the gray variety are generally regarded as inferior to the other two. This crop is grown to best advantage in climates where the nights are cool and moist. It matures more quickly than any other grain crop and is remarkably free from disease. The yield varies from ten to forty bushels an acre. Buckwheat does not seem to draw plant food heavily from the soil and can be grown on the same land from year to year.

In fertilizing buckwheat land, green manures and rich nitrogenous fertilizers should be avoided. These cause such a luxuriant growth that the stalks lodge badly.

The time of seeding will have to be settled by the height of the land and by the climate. In northern climates and in high altitudes the seeding is generally done in May or June. In southern climates and in low altitudes the planting may wait until July or August. The plant usually matures in about seventy days. It cannot stand warm weather at blooming-time, and must always be planted so that it may escape warm weather in its blooming period and cold weather in its maturing season. The seeds are commonly broadcasted at an average rate of four pecks to the acre. If the land is loose and pulverized, it should be rolled.

Buckwheat ripens unevenly and will continue to bloom until frost. Harvesting usually begins just after the first crop of seeds have matured. To keep the grains from shattering, the harvesting is best done during damp or cloudy days or early in the morning while the dew is still on the grain. The grain should be threshed as soon as it is dry enough to go through the thresher.

Buckwheat is grown largely for table use. The grain is crushed into a dark flour that makes most palatable breakfast cakes. The grain, especially when mixed with corn, is becoming popular for poultry food. The middlings, which are rich in fats and protein, are prized for dairy cows.

SECTION XLVIII. RICE

The United States produces only about one half of the rice that it consumes. There is no satisfactory reason for our not raising more of this staple crop, for five great states along the Gulf of Mexico are well adapted to its culture.

There are two distinct kinds of rice, upland rice and lowland rice. Upland rice demands in general the same methods of culture that are required by other cereals, for example, oats or wheat. The growing of lowland rice is considerably more difficult and includes the necessity of flooding the fields with water at proper times.

A stiff, half-clay soil with some loam is best suited to this crop. The soil should have a clay subsoil to retain water and to give stiffness enough to allow the use of harvesting-machinery. Some good rice soils are so stiff that they must be flooded to soften them enough to admit of plowing. Plow deeply to give the roots ample feeding-space. Good tillage, which is too often neglected, is valuable.

Careful seed-selection is perhaps even more needed for rice than for any other crop. Consumers want kernels of the same size. Be sure that your seed is free from red rice and other weeds. Drilling is much better than broadcasting, as it secures a more even distribution of the seed.

The notion generally prevails that flooding returns to the soil the needed fertility. This may be true if the flooding-water deposits much silt, but if the water be clear it is untrue, and fertilizers or leguminous crops are needed to keep up fertility. Cowpeas replace the lost soil-elements and keep down weeds, grasses, and red rice.

Red rice is a weed close kin to rice, but the seed of one will not produce the other. Do not allow it to get mixed and sowed with your rice seed or to go to seed in your field.

SECTION XLIX. THE TIMBER CROP

Forest trees are not usually regarded as a crop, but they are certainly one of the most important crops. We should accustom ourselves to look on our trees as needing and as deserving the same care and thought that we give to our other field crops. The total number of acres given to the growth of forest trees is still enormous, but we should each year add to this acreage.

Unfortunately very few forests are so managed as to add yearly to their value and to preserve a model stand of trees. Axmen generally fell the great trees without thought of the young trees that should at once begin to fill the places left vacant by the fallen giants. Owners rarely study their woodlands to be sure that the trees are thick enough, or to find out whether the saplings are ruinously crowding one another. Disease is often allowed to slip in unchecked. Old trees stand long after they have outlived their usefulness.

The farm wood-lot, too, is often neglected. As forests are being swept away, fuel is of course becoming scarcer and more costly. Every farmer ought to plant trees enough on his waste land to make sure of a constant supply of fuel. The land saved for the wood-lot should be selected from land unfit for cultivation. Steep hillsides, rocky slopes, ravines, banks of streams--these can, without much expense or labor, be set in trees and insure a never-ending fuel supply.

The most common enemies of the forest crop are:

First, forest fires. The waste from forest fires in the United States is most startling. Many of these fires are the result of carelessness or ignorance. Most of the states have made or are now making laws to prevent and to control such fires.

Second, fungous diseases. The timber loss from these diseases is exceedingly great.

Third, insects of many kinds prey on the trees. Some strip all the leaves from the branches. Others bore into the roots, trunk, or branches. Some lead to a slow death; others are more quickly fatal.

Fourth, improper grazing. Turning animals into young woods may lead to serious loss. The animals frequently ruin young trees by eating all the foliage. Hogs often unearth and consume most of the seeds needed for a good growth.

The handling of forests is a business just as the growing of corn is a business. In old forests, dead and dying trees should be cut. Trees that occupy space and yet have little commercial value should give way to more valuable trees. A quick-growing tree, if it is equally desirable, should be preferred to a slow grower. An even distribution of the trees should be secured.

In all there are about five hundred species of trees which are natives of the United States. Probably not over seventy of these are desirable for forests. In selecting trees to plant or to allow to grow from their own seeding, pick those that make a quick growth, that have a steady market value, and that suit the soil, the place of growth, and the climate.

SECTION L. THE FARM GARDEN

Every farmer needs a garden in which to grow not only vegetables but small fruits for the home table.

The garden should always be within convenient distance of the farmhouse. If possible, the spot selected should have a soil of mixed loam and clay. Every foot of soil in the garden should be made rich and mellow by manure and cultivation. The worst soils for the home garden are light, sandy soils, or stiff, clayey soils; but any soil, by judicious and intelligent culture, can be made suitable.

In laying out the garden we should bear in mind that hand labor is the most expensive kind of labor. Hence we should not, as is commonly done, lay off the garden spot in the form of a square, but we should mark off for our purpose a long, narrow piece of land, so that the cultivating tools may all be conveniently drawn by a horse or a mule. The use of the plow and the horse cultivator enables the cultivation of the garden to be done quickly, easily, and cheaply.

Each vegetable or fruit should be planted in rows, and not in little patches. Beginning with one side of the garden the following plan of arrangement is simple and complete: two rows to corn for table use; two to cabbages, beets, radishes, and eggplants; two to onions, peas, and beans; two to oyster-plants, okra, parsley, and turnips; two to tomatoes; then four on the other side can be used for strawberries, blackberries, raspberries, currants, and gooseberries.

The garden, when so arranged, can be tilled in the spring and tended throughout the growing season with little labor and little loss of time. In return for this odd-hour work, the farmer's family will have throughout the year an abundance of fresh, palatable, and health-giving vegetables and small fruits.

The keynote of successful gardening is to stir the soil. Stir it often with four objects in view:

1. To destroy weeds.

2. To let air enter the soil.

3. To enrich the soil by the action of the air.

4. To retain the moisture by preventing its evaporation.

corn corn

cabbage beets radishes cabbage beets eggplants

onions peas beans onions peas beans

oyster-plants okra parsley parsnips oyster-plants okra parsley parsnips

tomatoes tomatoes

strawberries currants raspberries blackberries strawberries currants raspberries blackberries strawberries currants raspberries blackberries strawberries currants raspberries blackberries

CHAPTER IX

FEED STUFFS

SECTION LI. GRASSES

Under usual conditions no farmer expects to grow live stock successfully and economically without setting apart a large part of his land for the growth of mowing and pasture crops. Therefore to the grower of stock the management of grass crops is all-important.

In planting either for a meadow or for a pasture, the farmer should mix different varieties of grass seeds. Nature mixes them when she plants, and Nature is always a trustworthy teacher.

In planting for a pasture the aim should be to sow such seeds as will give green grass from early spring to latest fall. In seeding for a meadow such varieties should be sowed together as ripen about the same time.

Even in those sections of the country where it grows sparingly and where it is easily crowded out, clover should be mixed with all grasses sowed, for it leaves in the soil a wealth of plant food for the grasses coming after it to feed on. Nearly every part of our country has some clover that experience shows to be exactly suited to its soil and climate. Study these clovers carefully and mix them with your grass seed.

The reason for mixing clover and grass is at once seen. The true grasses, so far as science now shows, get all their nitrogen from the soil; hence they more or less exhaust the soil. But, as several times explained in this book, the clovers are legumes, and all legumes are able by means of the bacteria that live on their roots to use the free nitrogen of the air. Hence without cost to the farmer these clovers help the soil to feed their neighbors, the true grasses. For this reason some light perennial legume should always be added to grass seed.

It is not possible for grasses to do well in a soil that is full of weeds. For this reason it is always best to sow grass in fields from which cultivated crops have just been taken. Soil which is to have grass sowed in it should have its particles pressed together. The small grass seeds cannot take root and grow well in land that has just been plowed and which, consequently, has its particles loose and comparatively far apart. On the other hand, land from which a crop of corn or cotton has just been harvested is in a compact condition. The soil particles are pressed well together. Such land when mellowed by harrowing makes a splendid bed for grass seeds. A firm soil draws moisture up to the seeds, while a mellow soil acts as a blanket to keep moisture from wasting into the air, and at the same time allows the heated air to circulate in the soil.

In case land has to be plowed for grass-seeding, the plowing should be done as far as possible in advance of the seeding. Then the plowed land should be harrowed several times to get the land in a soft, mellow condition.

If the seed-bed be carefully prepared, little work on the ground is necessary after the seeds are sowed. One light harrowing is sufficient to cover the broadcast seeds. This harrowing should always be done as soon as the seeds are scattered, for if there be moisture in the soil the tiny seeds will soon sprout, and if the harrowing be done after germination is somewhat advanced, the tender grass plants will be injured.

There are many kinds of pasture and meadow grasses. In New England, timothy, red clover, and redtop are generally used for the mowing crop. For permanent pasture, in addition to those mentioned, there should be added white clover and either Kentucky or Canadian blue grass. In the Southern states a good meadow or pasture can be made of orchard grass, red clover, and redtop. For a permanent pasture in the South, Japan clover, Bermuda, and such other local grasses as have been found to adapt themselves readily to the climate should be added. In the Middle States temporary meadows and pastures are generally made of timothy and red clover, while for permanent pastures white clover and blue grass thrive well. In the more western states the grasses previously suggested are readily at home. Alfalfa is proving its adaptability to nearly all sections and climates, and is in many respects the most promising grass crop of America.

It hardly ever pays to pasture meadows, except slightly, the first season, and then only when the soil is dry. It is also poor policy to pasture any kind of grass land early in the spring when the soil is wet, because the tramping of animals crushes and destroys the crowns of the plants. After the first year the sward becomes thicker and tougher, and the grass is not at all injured if it is grazed wisely.

The state of maturity at which grass should be harvested to make hay of the best quality varies somewhat with the different grasses and with the use which is to be made of the hay. Generally speaking, it is a good rule to cut grass for hay just as it is beginning to bloom or just after the bloom has fallen. All grasses become less palatable to stock as they mature and form seed. If grass be allowed to go to seed, most of the nutrition in the stalk is used to form the seed.

Hence a good deal of food is lost by waiting to cut hay until the seeds are formed.

Pasture lands and meadow lands are often greatly improved by replowing and harrowing in order to break up the turf that forms and to admit air more freely into the soil. The plant-roots that are destroyed by the plowing or harrowing make quickly available plant food by their decay, and the physical improvement of the soil leads to a thicker and better stand. In the older sections of the country commercial fertilizer can be used to advantage in producing hay and pasturage. If, however, clover has just been grown on grass land or if it is growing well with the grass, there is no need to add nitrogen. If the grass seems to lack sufficient nourishment, add phosphoric acid and potash.

However, grass not grown in company with clover often needs dried blood, nitrate of soda, or some other nitrogen-supplying agent. Of course it is understood that no better fertilizer can be applied to grass than barnyard manure.

SECTION LII. LEGUMES

Often land which was once thought excellent is left to grow up in weeds. The owner says that the land is worn out, and that it will not pay to plant it. What does "worn out" mean? Simply that constant cropping has used up the plant food in the land. Therefore, plants on worn-out land are too nearly starved to yield bountifully. Such wearing out is so easily prevented that no owner ought ever to allow his land to become poverty-stricken. But in case this misfortune has happened, how can the land be again made fertile?

On page 24 you learned that phosphoric acid, potash, and nitrogen are the foods most needed by plants. "Worn out," then, to put it in another way, usually means that a soil has been robbed of one of these plant necessities, or of two or of all three. To make the land once more fruitful it is necessary to restore the missing food or foods. How can this be done? Two of these plant foods, namely, phosphoric acid and potash, are minerals. If either of these is lacking, it can be supplied only by putting on the land some fertilizer containing the missing food. Fortunately, however, nitrogen, the most costly of the plant foods, can be readily and cheaply returned to poor land.

As explained on page 32 the leguminous crops have the power of drawing nitrogen from the air and, by means of their root-tubercles, of storing it in the soil. Hence by growing these crops on poor land the expensive nitrogen is quickly restored to the soil, and only the two cheaper plant foods need be bought. How important it is then to grow these leguminous plants! Every farmer should so rotate his planting that at least once every two or three years a crop of legumes may add to the fruitfulness of his fields.

Moreover these crops help land in another way. They send a multitude of roots deep into the ground. These roots loosen and pulverize the soil, and their decay, at the end of the growing season, leaves much humus in the soil. Land will rarely become worn out if legumes are regularly and wisely grown.

From the fact that they do well in so many different sections and in so many different climates, the following are the most useful legumes: alfalfa, clovers, cowpeas, vetches, and soy beans.

=Alfalfa.= Alfalfa is primarily a hay crop. It thrives in the Far West, in the Middle West, in the North, and in the South. In fact, it will do well wherever the soil is rich, moist, deep, and underlaid by an open subsoil. The vast areas given to this valuable crop are yearly increasing in every section of the United States. Alfalfa, however, unlike the cowpea, does not take to poor land. For its cultivation, therefore, good fertile land that is moist but not water-soaked should be selected.

Good farmers are partial to alfalfa for three reasons. First, it yields a heavy crop of forage or hay. Second, being a legume, it improves the soil. Third, one seeding lasts a long time. This length of life may, however, be destroyed by pasturing or abusing the alfalfa.

Alfalfa is different from most plants in this respect: the soil in which it grows must have certain kinds of bacteria in it. These cause the growth of tubercles on the roots. These bacteria, however, are not always present in land that has not been planted in alfalfa. Hence if this plant is to be grown successfully these helpful bacteria must sometimes be supplied artificially.

There are two very easy ways of supplying the germs. First, fine soil from an alfalfa field may be scattered broadcast over the fields to be seeded. Second, a small mass of alfalfa tubercle germs may be put into a liquid containing proper food to make these germs multiply and grow; then the seeds to be planted are soaked in this liquid in order that the germs may fasten on the seeds.

Before the seeds are sowed the soil should be mellowed. Over this well-prepared land about twenty pounds of seed to the acre should be scattered. The seed may be scattered by hand or by a seed-sower. Cover with a light harrow. The time of planting varies somewhat with the climate. Except where the winters are too severe the seed may be sowed either in the spring or in the fall. In the South sow only in the fall.

During the first season one mowing, perhaps more, is necessary to insure a good stand and also to keep down the weeds. When the first blossoms appear in the early summer, it is time to start the mower. After this the alfalfa should be cut every two, three, or four weeks. The number of times depends on the rapidity of growth.

This crop rarely makes a good yield the first year, but if a good stand be secured, the yield steadily increases. After a good stand has been secured, a top-dressing of either commercial fertilizer or stable manure will be very helpful. An occasional cutting-up of the sod with a disk harrow does much good.

=Clovers.= The different kinds of clovers will sometimes grow on hard or poor soil, but they do far better if the soil is enriched and properly prepared before the seed is sowed. In many parts of our country it has been the practice for generations to sow clover seed with some of the grain crops. Barley, wheat, oats, and rye are the crops with which clover is usually planted, but many good farmers now prefer to sow the seed only with other grass seed. Circumstances must largely determine the manner of seeding.

Crimson clover, which is a winter legume, usually does best when seeded alone, although rye or some other grain often seems helpful to it. This kind of clover is an excellent crop with which to follow cotton or corn. It is most conveniently sowed at the last cultivation of these crops.

Common red clover, which is the standard clover over most of the country, is usually seeded with timothy or with orchard grass or with some other of the grasses. In sowing both crimson and red clover, about ten to fifteen pounds of seed for each acre are generally used.

To make good pastures, white and Japan clover are favorites. White clover does well in most parts of America, and Japan clover is especially valuable in warm Southern climates. Both will do well even when the soil is partly shaded, but they do best in land fully open to the sun.

Careful attention is required to cure clover hay well. The clover should always be cut before it forms seed. The best time to cut is when the plants are in full bloom.

Let the mower be started in the morning. Then a few hours later run over the field with the tedder. This will loosen the hay and let in air and sunshine. If the weather be fair let the hay lie until the next day, and then rake it into rows for further drying. After being raked, the hay may either be left in the rows for final curing or it may be put in cocks. If the weather be unsettled, it is best to cock the hay. Many farmers have cloth covers to protect the cocks and these often aid greatly in saving the hay crop in a rainy season. In case the hay is put in cocks, it should be opened for a final drying before it is housed.

=Cowpeas.= The cowpea is an excellent soil-enricher. It supplies more fertilizing material to turn into the soil, in a short time and at small cost, than any other crop. Moreover, by good tillage and by the use of a very small amount of fertilizer, the cowpea can be grown on land too poor to produce any other crop. Its roots go deep into the soil. Hence they gather plant food and moisture that shallow-rooted plants fail to reach. These qualities make it an invaluable help in bringing worn-out lands back to fertility.

The cowpea is a warm-weather legume. In the United States it succeeds best in the south and southwest. It has, however, in recent years been grown as far north as Massachusetts, New York, Ohio, Michigan, and Minnesota, but in these cold climates other legumes are more useful. Cowpeas should never be planted until all danger of frost is past. Some varieties make their full growth in two months; others need four months.

There are about two hundred varieties of cowpeas. These varieties differ in form, in the size of seed and of pod, in the color of seed and of pod, and in the time of ripening. They differ, too, in the manner of growth. Some grow erect; others sprawl on the ground. In selecting varieties it is well to choose those that grow straight up, those that are hardy, those that fruit early and abundantly, and those that hold their leaves. The variety selected for seed should also suit the land and the climate.

The cowpea will grow in almost any soil. It thrives best and yields most bountifully on well-drained sandy loams. The plant also does well on clay soils. On light, sandy soils a fairly good crop may be made, but on such soils, wilt and root-knot are dangerous foes. A warm, moist, well-pulverized seed-bed should always be provided. Few plants equal the cowpea in repaying careful preparation.

If this crop is grown for hay, the method of seeding and cultivating will differ somewhat from the method used when a seed crop is desired. When cowpeas are planted for hay the seeds should be drilled or broadcasted. If the seeds are small and the land somewhat rich, about four pecks should be sowed on each acre. If the seeds are comparatively large and the soil not so fertile, about six pecks should be sowed to the acre. It is safer to disk in the seeds when they are sowed broadcast than it is to rely on a harrow to cover them. In sowing merely for a hay crop, it is a good practice to mix sorghum, corn, soy beans, or millet with the cowpeas. The mixed hay is more easily harvested and more easily cured than unmixed cowpea hay. Shortly after seeding, it pays to run over the land lightly with a harrow or a weeder in order to break any crust that may form.

Mowing should begin as soon as the stalks and the pods have finished growing and some of the lower leaves have begun to turn yellow. An ordinary mower is perhaps the best machine for cutting the vines. If possible, select only a bright day for mowing and do not start the machine until the dew on the vines is dried. Allow the vines to remain as they fell from the mower till they are wilted; then rake them into windrows. The vines should generally stay in the windrows for two or three days and be turned on the last day. They should then be put in small, airy piles or piled around a stake that has crosspieces nailed to it. The drying vines should never be packed; air must circulate freely if good hay is to be made. As piling the vines around stakes is somewhat laborious, some growers watch the curing carefully and succeed in getting the vines dry enough to haul directly from the windrows to the barns. Never allow the vines to stay exposed to too much sunshine when they are first cut. If the sun strikes them too strongly, the leaves will become brittle and shatter when they are moved.

When cowpeas are grown for their pods to ripen, the seeds should be planted in rows about a yard apart. From two to three pecks of seeds to an acre should be sufficient. The growing plants should be cultivated two or three times with a good cultivator. Cowpeas were formerly gathered by hand, but such a method is of course slow and expensive. Pickers are now commonly used.

Some farmers use the cowpea crop only as a soil-enricher. Hence they neither gather the seeds nor cut the hay, but plow the whole crop into the soil. There is an average of about forty-seven pounds of nitrogen in each ton of cowpea vines. Most of this valuable nitrogen is drawn by the plants from the air. This amount of nitrogen is equal to that contained in 9500 pounds of stable manure. In addition each ton of cowpea vines contains ten pounds of phosphoric acid and twenty-nine pounds of potash.

There is danger in plowing into the soil at one time any bountiful green crop like cowpeas. As already explained on page 10, a process called capillarity enables moisture to rise in the soil as plants need it. Now if a heavy cowpea crop or any other similar crop be at one plowing turned into the soil, the soil particles will be so separated as to destroy capillarity. Too much vegetation turned under at once may also, if the weather be warm, cause fermentation to set in and "sour the land." Both of these troubles may be avoided by cutting up the vines with a disk harrow or other implement before covering them.

The custom of planting cowpeas between the rows at the last working of corn is a good one, and wherever the climate permits this custom should be followed.

=Vetches.= The vetches have been rapidly growing in favor for some years. Stock eat vetch hay greedily, and this hay increases the flow of milk in dairy animals and helps to keep animals fat and sleek. Only two species of vetch are widely grown. These are the tare, or spring vetch, and the winter, or hairy, vetch. Spring vetch is grown in comparatively few sections of our country. It is, however, grown widely in England and northern continental Europe. What we say here will be confined to hairy vetch.

After a soil has been supplied with the germs needed by this plant, the hairy vetch is productive on many different kinds of soil. The plant is most vigorous on fertile loams. By good tillage and proper fertilization it may be forced to grow rather bountifully on poor sandy and clay loams. Acid or wet soils are not suited to vetch. Lands that are too poor to produce clovers will frequently yield fair crops of vetch. If this is borne in mind, many poor soils may be wonderfully improved by growing on them this valuable legume.

Vetch needs a fine well-compacted seed-bed, but it is often sowed with good results on stubble lands and between cotton and corn rows, where it is covered by a cultivator or a weeder.

The seeds of the vetch are costly and are brought chiefly from Germany, where this crop is much prized. The pods ripen so irregularly that they have to be picked by hand.

In northern climates early spring sowing is found most satisfactory. In southern climates the seeding is best done in the late summer or early fall. As the vetch vines have a tendency to trail on the ground, it is wisest to plant with the vetch some crop like oats, barley, rye, or wheat. These plants will support the vetch and keep its vines from being injured by falling on the ground. Do not use rye with vetch in the South. It ripens too early to be of much assistance. If sowed with oats the seeding should be at the rate of about twenty or thirty pounds of vetch and about one and a half or two bushels of oats to the acre. Vetch is covered in the same way as wheat and rye.

Few crops enrich soil more rapidly than vetch if the whole plant is turned in. It of course adds nitrogen to the soil and at the same time supplies the soil with a large amount of organic matter to decay and change to humus. As the crop grows during the winter, it makes an excellent cover to prevent washing. Many orchard-growers of the Northwest find vetch the best winter crop for the orchards as well as for the fields.

=Soy, or Soja, Bean.= In China and Japan the soy bean is grown largely as food for man. In the United States it is used as a forage plant and as a soil-improver. It bids fair to become one of the most popular of the legumes. Like the cowpea, this bean is at home only in a warm climate. Some of the early-ripening varieties have, however, been planted with fair success in cold climates.

While there are a large number of varieties of the soy bean, only about a dozen are commonly grown. They differ mainly in the color, size, and shape of the seeds, and in the time needed for ripening. Some of the varieties are more hairy than others.

Soy beans may take many places in good crop-rotations, but they are unusually valuable in short rotations with small grains. The grains can be cut in time for the beans to follow them, and in turn the beans can be harvested in the early fall and make way for another grain crop.

It should always be remembered that soy beans will not thrive unless the land on which they are to grow is already supplied, or is supplied at the time of sowing, with bean bacteria.

The plant will grow on many different kinds of soil, but it needs a richer soil than the cowpea does. As the crop can gather most of its own nitrogen, it generally requires only the addition of phosphoric acid and potash for its growth on poor land. When the first crop is seeded, apply to each acre four hundred pounds of a fertilizing mixture which contains about ten per cent of phosphoric acid, four per cent of potash, and from one to two per cent of nitrogen.

If the crop is planted for hay or for grazing, mellow the ground well, and then broadcast or drill in closely about one and a half bushels of seed to each acre. Cover from one to two inches deep, but never allow a crust to form over the seed, for the plant cannot break through a crust well. When the beans are planted for seeds, a half bushel of seed to the acre is usually sufficient. The plants should stand in the rows from four to six inches apart, and the rows should be from thirty to forty inches from one another. Never plant until the sun has thoroughly warmed the land. The bean may be sowed, however, earlier than cowpeas. A most convenient time is just after corn is planted. The rows should be cultivated often enough to keep out weeds and grass and to keep a good dust mulch, but the cultivation must be shallow.

As soy beans are grown for hay and also for seed, the harvesting will, as with the other legumes, be controlled by the purpose for which the crop was planted. In harvesting for a hay crop it is desirable to cut the beans after the pods are well formed but before they are fully grown. If the cutting is delayed until the pods are ripe, the fruit will shatter badly. There is a loss, too, in the food value of the stems if the cutting is late. The ordinary mowing-machine with a rake attached is generally the machine used for cutting the stalks. The leaves should be most carefully preserved, for they contain much nourishment for stock.

Whenever the beans are grown for seeds, harvesting should begin when three fourths of the leaves have fallen and most of the pods are ripe. Do not wait, however, until the pods are so dry that they have begun to split and drop their seeds. A slight amount of dampness on the plants aids the cutting. The threshing may be done with a flail, with pea-hullers, or with a grain-threshing machine.

The beans produce more seed to the acre than cowpeas do. Forty bushels is a high yield. The average yield is between twenty and thirty bushels.

DESCRIPTIVE TABLE

ADAPTATION AS Crop FOOD FOR ANIMALS LIFE REMARKS

Alfalfa Hay Perennial All animals like it; hogs eat it even when it is dry. Red clover Hay and pasture Perennial Best of the clovers for hay. Alsike clover Hay and pasture Perennial Seeds itself for twenty years. This clover is a great favorite with bees. Mammoth clover Hay and pasture Perennial Best for green manure. White clover Pasture Perennial Excellent for lawns and bees. Japan clover Pasture Perennial Excellent for forest and old soils. Cowpea Hay and grain Annual Used for hay, green manure, and pastures. Soy bean Hay and grain Annual Often put in silo with corn. Vetches Hay and soiling Annual Pasture for sheep and swine. With cereals it makes excellent hay and soiling-food.

CHAPTER X

DOMESTIC ANIMALS

The progress that a nation is making can with reasonable accuracy be measured by the kind of live stock it raises. The general rule is, poor stock, poor people. All the prosperous nations of the globe, especially the grain-growing nations, get a large share of their wealth from raising improved stock. The stock bred by these nations is now, however, very different from the stock raised by the same nations years ago. As soon as man began to progress in the art of agriculture he became dissatisfied with inferior stock. He therefore bent his energies to raise the standard of excellence in domestic animals.

By slow stages of animal improvement the ugly, thin-flanked wild boar of early times has been transformed into the sleek Berkshire or the well-rounded Poland-China. In the same manner the wild sheep of the Old World have been developed into wool and mutton breeds of the finest excellence. By constant care, attention, and selection the thin, long-legged wild ox has been bred into the bounteous milk-producing Jerseys and Holsteins or into the Shorthorn mountains of flesh. From the small, bony, coarse, and shaggy horse of ancient times have descended the heavy Norman, or Percheron, draft horse and the fleet Arab courser.

The matter of meat-production is one of vital importance to the human race, for animal food must always supply a large part of man's ration.

Live stock of various kinds consume the coarser foods, like the grasses, hays, and grains, which man cannot use. As a result of this consumption they store in their bodies the exact substances required for building up the tissues of man's body.

When the animal is used by man for food, one class of foods stored away in the animal's body produces muscle; another produces fat, heat, and energy. The food furnished by the slaughter of animals seems necessary to the full development of man. It is true that the flesh of an animal will not support human life so long as would the grain that the animal ate while growing, but it is also true that animal food does not require so much of man's force to digest it. Hence the use of meat forces a part of man's life-struggle on the lower animal.

When men feed grain to stock, the animals receive in return power and food in their most available forms. Men strengthen the animal that they themselves may be strengthened. One of the great questions, then, for the stock-grower's consideration is how to make the least amount of food fed to animals produce the most power and flesh.

SECTION LIII. HORSES

While we have a great many kinds of horses in America, horses are not natives of this country. Just where wild horses were first tamed and used is not certainly known. It is believed that in early ages the horse was a much smaller animal than it now is, and that it gradually attained its present size. Where food was abundant and nutritious and the climate mild and healthful, the early horses developed large frames and heavy limbs and muscles; on the other hand, where food was scarce and the climate cold and bleak, the animals remained as dwarfed as the ponies of the Shetland Islands.

One of the first records concerning the horse is found in Genesis xlix, 17, where Jacob speaks of "an adder that biteth the horse heels." Pharaoh took "six hundred chosen chariots" and "with all the horses and chariots" pursued the Israelites. The Greeks at first drove the horse fastened to a rude chariot; later they rode on its back, learning to manage the animal with voice or switch and without either saddle or bridle. This thinking people soon invented the snaffle bit, and both rode and drove with its aid. The curb bit was a Roman invention. Shoeing was not practiced by either Greeks or Romans. Saddles and harnesses were at first made of skins and sometimes of cloth.

Among the Tartars of middle and northern Asia and also among some other nations, mare's milk and the flesh of the horse are used for food. Old and otherwise worthless horses are regularly fattened for the meat markets of France and Germany. Various uses are made of the different parts of a horse's body. The mane and tail are used in the manufacture of mattresses, and also furnish a haircloth for upholstering; the skin is tanned into leather; the hoofs are used for glue, and the bones for making fertilizer.

Climate, food, and natural surroundings have all aided in producing changes in the horse's form, size, and appearance. The varying circumstances under which horses have been raised have given rise to the different breeds. In addition, the masters' needs had much to do in developing the type of horses wanted. Some masters desired work horses, and kept the heavy, muscular, stout-limbed animals; others desired riding and driving horses, so they saved for their use the light-limbed, angular horses that had endurance and mettle. The following table gives some of the different breeds and the places of their development:

I. Draft, or Heavy, Breeds

1. Percheron, from the province of Perche, France. 2. French Draft, developed in France. 3. Belgian Draft, developed by Belgian farmers. 4. Clydesdale, the draft horse of Scotland. 5. Suffolk Punch, from the eastern part of England. 6. English Shire, also from the eastern part of England.

II. Carriage, or Coach, Breeds

1. Cleveland Bay, developed in England. 2. French Coach, the gentleman's horse of France. 3. German Coach, from Germany. 4. Oldenburg Coach, Oldenburg, Germany. 5. Hackney, the English high-stepper.

III. Light, or Roadster, Breeds

1. American Trotter, developed in America. 2. Thoroughbred, the English running horse. 3. American Saddle Horse, from Kentucky and Virginia.

There is a marked difference in the form and type of these horses, and on this difference their usefulness depends.

The draft breeds have short legs, and hence their bodies are comparatively close to the ground. The depth of the body should be about the same as the length of leg. All draft horses should have upright shoulders, so as to provide an easy support for the collar. The hock should be wide, so that the animal shall have great leverage of muscle for pulling. A horse having a narrow hock is not able to draw a heavy load and is easily exhausted and liable to curb-diseases (see Figs. 242 and 243).

The legs of all kinds of horses should be straight; a line dropped from the point of the shoulder to the ground should divide the knees, canon, fetlock, and foot into two equal parts. When the animal is formed in this way the feet have room to be straight and square, with just the breadth of a hoof between them (Fig. 241).

Roadsters are lighter in bone and less heavily muscled; their legs are longer than those of the draft horses and, as horsemen say, more "daylight" can be seen under the body. The neck is long and thin, but fits nicely into the shoulders. The shoulders are sloping and long and give the roadster ability to reach well out in his stride. The head is set gracefully on the neck and should be carried with ease and erectness.

Every man who is to deal with horses ought to become, by observation and study, an expert judge of forms, qualities, types, defects, and excellences.

The horse's foot makes an interesting study. The horny outside protects the foot from mud, ice, and stones. Inside the hoof are the bones and gristle that serve as cushions to diminish the shock received while walking or running on hard roads or streets. When shoeing the horse the frog should not be touched with the knife. It is very seldom that any cutting need be done. Many blacksmiths do not know this and often greatly injure the foot.

Since the horse has but a small stomach, the food given should not be too bulky. In proportion to the horse's size, its grain ration should be larger than that of other animals. Draft horses and mules, however, can be fed a more bulky ration than other horses, because they have larger stomachs and consequently have more room to store food.

The horse should be groomed every day. This keeps the pores of the skin open and the hair bright and glossy. When horses are working hard, the harness should be removed during the noon hour. During the cool seasons of the year, whenever a horse is wet with sweat, it should on stopping work, or when standing for awhile, be blanketed, for the animal is as liable as man to get cold in a draft or from moisture evaporating rapidly from its skin.

EXERCISE

If the pupil will take an ordinary tape measure, he can make some measurements of the horse that will be very interesting as well as profitable. Let him measure:

1. The height of the horse at the withers, 1 to 1. 2. The height of the horse at croup, 2 to 2. 3. Length of shoulder, 1 to 3. 4. Length of back, 4. 5. Length of head, 5. 6. Depth of body, 6 to 6. 7. Daylight under body, 7 to 7. 8. Distance from point of shoulder to quarter, 3 to 3. 9. Width of forehead. 10. Width between hips.

NOTE. Many interesting comparisons can be made (1) by measuring several horses; (2) by studying the proportion between parts of the same horse.

PROPORTIONS OF A HORSE

1. How many times longer is the body than the head? Do you get the same result from different horses?

2. How does the height at the withers compare with the height at the croup?

3. How do these compare with the distance from quarter to shoulder?

4. How does the length of the head compare with the thickness of the body and with the open space, or "daylight," under the body?

SECTION LIV. CATTLE

All farm animals were once called cattle; now this term applies only to beef and dairy animals--neat cattle.

Our improved breeds are descended from the wild ox of Europe and Asia, and have attained their size and usefulness by care, food, and selection. The uses of cattle are so familiar that we need scarcely mention them. Their flesh is a part of man's daily food; their milk, cream, butter, and cheese are on most tables; their hides go to make leather, and their hair for plaster; their hoofs are used for glue, and their bones for fertilizers, ornaments, buttons, and many other purposes.

There are two main classes of cattle--beef breeds and dairy breeds. The principal breeds of each class are as follows:

I. Beef Breeds

1. Aberdeen-Angus, bred in Scotland, and often called doddies. 2. Galloway, from Scotland. 3. Shorthorn, an English breed of cattle. 4. Hereford, also an English breed. 5. Sussex, from the county of Sussex, England.

II. Dairy Breeds

1. Jersey, from the Isle of Jersey. 2. Guernsey, from the Isle of Guernsey. 3. Ayrshire, from Scotland. 4. Holstein-Frisian, from Holland and Denmark. 5. Brown Swiss, from Switzerland.

Other breeds of cattle are Devon, Dutch Belted, Red-Polled, Kerry, and West Highland.

In general structure there is a marked difference between the beef and dairy breeds. This is shown in Figs. 248, 249. The beef cow is square, full over the back and loins, and straight in the back. The hips are covered evenly with flesh, the legs full and thick, the under line, or stomach line, parallel to the back line, and the neck full and short. The eye should be bright, the face short, the bones of fine texture, and the skin soft and pliable.

The dairy cow is widely different from the beef cow. She shows a decided wedge shape when you look at her from front, side, or rear. The back line is crooked, the hip bones and tail bone are prominent, the thighs thin and poorly fleshed; there is no breadth to the back, as in the beef cow, and little flesh covers the shoulders; the neck is long and thin.

The udder of the dairy cow is most important. It should be full but not fleshy, be well attached behind, and extend well forward. The larger the udder the more milk will be given.

The skin of the dairy cow, like that of the beef breeds, should be soft and pliable and the bones fine-textured.

=The Dairy Type.= Because of lack of flesh on the back, loins, and thighs, the cow of the dairy type is not profitably raised for beef, nor is the beef so good as that of the beef types. This is because in the dairy-animal food goes to produce milk rather than beef. In the same way the beef cow gives little milk, since her food goes rather to fat than to milk. For the same reasons that you do not expect a plow horse to win on the race track, you do not expect a cow of the beef type to win premiums as a milker.

"Scrub" cattle are not profitable. They mature slowly and consequently consume much food before they are able to give any return for it. Even when fattened, the fat and lean portions are not evenly distributed, and "choice cuts" are few and small.

By far the cheapest method of securing a healthy and profitable herd of dairy or beef cattle is to save only the calves whose sires are pure-bred animals and whose mothers are native cows. In this way farmers of even little means can soon build up an excellent herd.

=Improving Cattle.= The fact that it is not possible for every farmer to possess pure-bred cattle is no reason why he should not improve the stock he has. He can do this by using pure-bred sires that possess the qualities most to be desired. Scrub stock can be quickly improved by the continuous use of good sires. It is never wise to use grade, or cross-bred, sires, since the best qualities are not fixed in them.

Moreover, it is possible for every farmer to determine exactly the producing-power of his dairy cows. When the cows are milked, the milk should be weighed and a record kept. If this be done, it will be found that some cows produce as much as five hundred, and some as much as ten hundred, gallons a year, while others produce not more than two or three hundred gallons. If a farmer kills or sells his poor cows and keeps his best ones, he will soon have a herd of only heavy milkers. Ask your father to try this plan. Read everything you can find about taking care of cows and improving them, and then start a herd of your own.

=Conclusions.= (1) A cow with a tendency to get fat is not profitable for the dairy. (2) A thin, open, angular cow will make expensive beef. (3) "The sire is half the herd." This means that a good sire is necessary to improve a herd of cattle. The improvement from scrubs upward is as follows: the first generation is one-half pure; the second is three-fourths pure; the third is seven-eighths pure; the fourth is fifteen-sixteenths pure, etc. (4) By keeping a record of the quantity and quality of milk each cow gives you can tell which are profitable to raise from and which are not. (5) Good food, clean water, kindness, and care are necessary to successful cattle-raising.

The ownership of a well-bred animal usually arouses so much pride in the owner that the animal receives all the care that it merits. The watchful care given to such an animal leads to more thought of the other animals on the farm, and often brings about the upbuilding of an entire herd.

SECTION LV. SHEEP

The sheep was perhaps the first animal domesticated by man, and to-day the domesticated sheep is found wherever man lives. It is found domesticated or wild in almost every climate, and finds means to thrive where other animals can scarcely live; it provides man with meat and clothing, and is one of the most profitable and most easily cared-for of animals.

Sheep increase so rapidly, mature at such an early age, and have flesh so wholesome for food that nearly every farm should have its flock. Another consideration that may be urged in favor of sheep-raising is that sheep improve the land on which they are pastured.

Sheep are docile and easily handled, and they live on a greater diversity of food and require less grain than any other kind of live stock. In mixed farming there is enough food wasted on most farms to maintain a small flock of sheep.

Sheep may be divided into three classes:

I. Fine-Wooled Breeds

1. American Merino. 2. Delaine Merino. 3. Rambouillets. 4. Hampshire Down. 5. Oxford Down. 6. Cheviot.

II. Medium-Wooled Breeds

1. Southdown. 2. Shropshire. 3. Horned Dorset.

III. Long-Wooled Breeds

1. Leicester. 2. Lincoln. 3. Cotswold.

The first group is grown principally for wool, and mutton is secondary; in the second group, mutton comes first and wool second; in the third group both are important considerations. Wool is nature's protection for the sheep. Have you ever opened the fleece and observed the clean skin in which the fibers grow? These fibers, or hairs, are so roughened that they push all dirt away from the skin toward the outside of the fleece.

Wool is valuable in proportion to the length and evenness of the fiber and the density of the fleece.

EXERCISE

1. How many pounds ought a fleece of wool to weigh? 2. Which makes the better clothing, coarse or fine wool? 3. Why are sheep washed before being sheared? 4. Does cold weather trouble sheep? wet weather?

SECTION LVI. SWINE

The wild boar is a native of Europe, Asia, and Africa. The wild hogs are the parents from which all our domestic breeds have sprung. In many parts of the world the wild boar is still found. These animals are active and powerful, and as they grow older are fierce and dangerous. In their wild state they seek moist, sandy, and well-wooded places, close to streams of water. Their favorite foods are fruits, grass, and roots, but when pressed by hunger they will eat snakes, worms, and even higher animals, like birds, fowls, and fish.

Man captured some of these wild animals, fed them abundant and nutritious food, accustomed them to domestic life, selected the best of them to raise from, and in the course of generations developed our present breeds of hogs. The main changes brought about in hogs were these: the legs became shorter, the snout and neck likewise shortened, the shoulders and hams increased their power to take on flesh, and the frame was strengthened to carry the added burden of flesh. As the animal grew heavier it roamed less widely, and as it grew accustomed to man its temper became less fierce.

Meat can be more cheaply obtained from hogs than from any other animal. When a hog is properly fed and cared for it will make the farmer more money in proportion to cost than any other animal on the farm.

The most profitable type of hog has short legs, small bones, straight back and under line, heavy hams, small well-dished head, and heavy shoulders. The scrub and "razorback" hogs are very unprofitable, and require an undue amount of food to produce a pound of gain. It requires two years to get the scrub to weigh what a well-bred pig will weigh when nine months old. Scrub hogs can be quickly changed in form and type by the use of a pure-bred sire.

A boy whose parents were too poor to send him to college once decided to make his own money and get an education. He bought a sow and began to raise pigs. He earned the food for the mother and her pigs. His hogs increased so rapidly that he had to work hard to keep them in food. By saving the money he received from the sale of his hogs he had enough to keep him two years in college. Suppose you try his plan, and let the hog show you how fast it can make money.

We have several breeds of swine. The important ones are:

I. Large Breeds

1. Chester White. 2. Improved Yorkshire. 3. Tamworth.

II. Medium Breeds

1. Berkshire. 2. Poland-China. 3. Duroc-Jersey. 4. Cheshire.

III. Small Breeds

1. Victoria. 2. Suffolk. 3. Essex. 4. Small Yorkshire.

Hogs will be most successfully raised when kept as little as possible in pens. They like the fields and the pasture grass, the open air and the sunshine. Almost any kind of food can be given them. Unlike other stock, they will devour greedily and tirelessly the richest feeding-stuffs.

The most desirable hog to raise is one that will produce a more or less even mixture of fat and lean. Where only corn is fed, the body becomes very fat and is not so desirable for food as when middlings, tankage, cowpeas, or soy beans are added as a part of the ration.

When hogs are kept in pens, cleanliness is most important, for only by cleanliness can disease be avoided.

SECTION LVII. FARM POULTRY

Our geese, ducks, turkeys, and domestic hens are all descendants of wild fowls, and are more or less similar to them in appearance.

The earliest recorded uses of fowls were for food, for fighting, and for sacrifice. To-day the domestic fowl has four well-defined uses--egg-production, meat-production, feather-production, and pest-destruction.

Hens of course produce most of our eggs. Some duck eggs are sold for table use. Goose and duck body-feathers bring good prices. As pest-destroyers turkeys and chickens are most useful. They eat large numbers of bugs and worms that are harmful to crops. A little proper attention would very largely increase the already handsome sum derived from our fowls. They need dry, warm, well-lighted, and tidily kept houses. They must have, if we want the best returns, an abundant supply of pure water and a variety of nutritious foods. In cold, rainy, or snowy weather they should have a sheltered yard, and in good weather should be allowed a range wide enough to give them exercise. Their bodies and their nests must be protected from every form of vermin.

For eggs, the Leghorn varieties are popular. Some hens of this breed have been known to lay more than two hundred eggs in a year. Specially cared-for flocks have averaged eleven or even twelve dozen eggs a year. Farm flocks of ordinary breeds average less than eight dozen. Other excellent egg breeds are the Spanish, Andalusian, and Minorca.

The principal so-called meat breeds are the Brahma, Cochin, and Langshan. These are very large, but rather slow-growing fowls, and are not noted as layers. They are far less popular in America, even as meat-producers, than the general-purpose breeds.

The Plymouth Rock, Wyandotte, Rhode Island Red, and Orpington are the leading general-purpose breeds. They are favorites because they are at once good-sized, good layers, tame, and good mothers. The chicks of these breeds are hardy and thrifty. In addition to these breeds, there are many so-called fancy breeds that are prized for their looks rather than for their value. Among these are the Hamburg, Polish, Sultan, Silkie, and the many Bantam breeds.

The leading duck breeds are the Pekin, Aylesbury, Indian Runner, Muscovy, Rouen, and Cayuga. The principal varieties of geese are the Toulouse, Emden, Chinese, and African.

Among the best breeds of turkeys are the Bronze, White Holland, Narragansett, Bourbon, Slate, and Buff.

Geese, ducks, and turkeys are not so generally raised as hens, but there is a constant demand at good prices for these fowls.

The varieties of the domestic hen are as follows:

I. Egg Breeds

1. Leghorn. 2. Minorca. 3. Spanish. 4. Blue Andalusian. 5. Anconas.

II. Meat Breeds

1. Brahma. 2. Cochin. 3. Langshan. 4. Dorking. 5. Cornish.

III. General-Purpose Breeds

1. Plymouth Rock. 2. Wyandotte. 3. Rhode Island Red. 4. Orpington.

IV. Fancy Breeds

1. Polish. 2. Game. 3. Sultan. 4. Bantam.

As the price of both eggs and fowls is steadily advancing, a great many people are now raising fowls by means of an incubator for hatching, and a brooder as a substitute for the mother hen.

The use of the incubator is extending each year and is now almost universal where any considerable number of chicks are to be hatched. Doubtless it will continue to be used wherever poultry-production is engaged in on a large scale.

The brooder is employed to take care of the chickens as soon as they leave the incubator.

SECTION LVIII. BEE CULTURE

Stock-raisers select breeds that are best adapted to their needs. Plant-growers exercise great care in their choice of plants, selecting for each planting those best suited to the conditions under which they are to be grown. Undoubtedly a larger yield of honey could be had each year if similar care were exercised in the selection of the breed of bees.

To prove this, one has only to compare the yield of two different kinds. The common East Indian honey bee rarely produces more than ten or twelve pounds to a hive, while the Cyprian bee, which is a most industrious worker, has a record of one thousand pounds in one season from a single colony. This bee, besides being industrious when honey material is plentiful, is also very persevering when such material is hard to find. The Cyprians have two other very desirable qualities. They stand the cold of winter well and stoutly defend their hives against robber bees and other enemies.

The Italian is another good bee. This variety was brought into the United States in 1860. While the yield from the Italian is somewhat less than from the Cyprian, the Italian bees produce a whiter comb and are a trifle more easily managed.

The common black or brown bee is found wild and domesticated throughout the country. When honey material is abundant, these bees equal the Italians in honey-production, but when the season is poor, they fall far short in the amount of honey produced.

The purchase of a good Cyprian or Italian hive will richly repay the buyer. Such a colony will cost more at the outset than an ordinary colony, but will soon pay for its higher cost by greater production.

A beehive in the spring contains one queen, several hundred drones, and from thirty-five to forty thousand workers. The duty of the queen is to lay all the eggs that are to hatch the future bees. This she does with untiring industry, often laying as many as four thousand in twenty-four hours.

The worker bees do all the work. Some of them visit the flowers, take up the nectar into the honey-sac, located in their abdomens, and carry it to the hive. They also gather pollen in basketlike cavities in their hind legs. Pollen and nectar are needed to prepare food for the young bees. In the hive other workers create a breeze by buzzing with their wings and produce heat by their activity-- all to cause the water to evaporate from the nectar and to convert it into honey before it is sealed up in the comb. After a successful day's gathering you may often hear these tireless workers buzzing till late into the night or even all through the night.

You know that the bees get nectar from the flowers of various plants. Some of the chief honey plants are alfalfa, buckwheat, horsemint, sourwood, white sage, wild pennyroyal, black gum, holly, chestnut, magnolia, and the tulip tree. The yield of honey may often be increased by providing special pasturage for the bees. The linden tree, for example, besides being ornamental and valuable for timber, produces a most bee-inviting flower. Vetch, clover, and most of the legumes and mints are valuable plants to furnish pasture for bees. Catnip may be cultivated for the bees and sold as an herb as well.

In spraying fruit trees to prevent disease you should always avoid spraying when the trees are in bloom, since the poison of the spray seriously endangers the lives of bees.

The eggs laid by the queen, if they are to produce workers, require about twenty-one days to bring forth the perfect bee. The newly hatched bee commences life as a nurse. When about ten days old it begins to try its wings in short flights, and a few days later it begins active work. The life of a worker bee in the busy season is only about six weeks. You may distinguish young exercising bees from real workers by the fact that they do not fly directly away on emerging from the hive, but circle around a bit in order to make sure that they can recognize home again, since they would receive no cordial welcome if they should attempt to enter another hive. They hesitate upon returning from even these short flights, to make sure that they are in front of their own door.

There are several kinds of enemies of the bee which all beekeepers should know. One of these is the robber bee, that is, a bee from another colony attempting to steal honey from the rightful owners, an attempt often resulting in frightful slaughter. Much robbery can be avoided by clean handling; that is, by leaving no honey about to cultivate a taste for stolen sweets. The bee moth is another serious enemy. The larva of the moth feeds on the wax. Keep the colonies of bees strong so that they may be able to overcome this moth.

Queenless or otherwise weak colonies should be protected by a narrow entrance that admits only one bee at a time, for such a pass may be easily guarded. Fig. 267 shows a good anti-robbery entrance which may be readily provided for every weak colony. Mice may be kept out by tin-lined entrances.

The widespread fear of the kingbird seems unfounded. He rarely eats anything but drones, and few of them. This is also true of the swallow. Toads, lizards, and spiders are, however, true enemies of the honeybee.

EXERCISE

Can you recognize drones, workers, and queens? Do bees usually limit their visits to one kind of blossom on any one trip? What effect has the kind of flower on the flavor of the honey produced? What kinds of flowers should the beekeeper provide for his bees? Is the kingbird really an enemy to the bee?

SECTION LIX. WHY WE FEED ANIMALS

In the first place, we give various kinds of feed stuffs to our animals that they may live. The heart beats all the time, the lungs contract and expand, digestion is taking place, the blood circulates through the body--something must supply force for these acts or the animal dies. This force is derived from food.

In the next place, food is required to keep the body warm. Food in this respect is fuel, and acts in the same way that wood or coal does in the stove. Our bodies are warm all the time, and they are kept warm by the food we eat at mealtime.

Then, in the third place, food is required to enable the body to enlarge--to grow. If you feed a colt just enough to keep it alive and warm, there will be no material present to enable it to grow; hence you must add enough food to form bone and flesh and muscle and hair and fat.

In the fourth place, we feed to produce strength for work. An animal poorly fed cannot do so much work at the plow or on the road as one that receives all the food needed.

Both food and the force produced by it result from the activity of plants. By means of sunlight and moisture a sprouting seed, taking out of the air and soil different elements, grows into a plant. Then, just as the plant feeds on the air and soil to get its growth, so the animal feeds on the plant, to get its growth. Hence, since our animals feed upon plants, we must find out what is in plants in order to know what animal food consists of.

Plants contain protein, carbohydrates, fat, mineral matter, water, and vitamins. You have seen protein compounds like the white of an egg, lean meat, or the gluten of wheat. The bodies of plants do not contain very much protein. On the other hand, all plant seeds contain a good deal of this substance. Animals make use of protein to form new blood, muscles, and organs. Because of the quality of protein, milk is the best food for children and young animals.

The protein in some foods is of poor quality. To insure a well-balanced supply of protein a variety in foods is desirable. Do not rely on a single kind of mill feed, but combine several kinds, such as cotton-seed meal, linseed meal, wheat bran and middlings, gluten, and similar grain by-products. Tankage for young pigs and meat scraps for chickens are high-grade proteins and are of animal origin.

It is no less important to get the necessary vitamins--those mysterious substances that keep the body healthy and promote growth and well-being. Scientists claim that many diseases are food-deficiency diseases--the body gets out of order because these peculiar vitamins are lacking in the food. Children require about one or two quarts of milk a day, fresh fruits, cereal breakfast foods, leafy vegetables as salads, and cooked vegetables.

Farm animals require the vitamins also. The legume pasture or hay, milk, grain concentrates when supplied in variety, pasture grass, and green forage crops are basic foods for farm animals. Very young animals should have milk also.

Let us next consider the carbohydrates. Sometimes the words starchy foods are used to describe the carbohydrates. You have long known forms of these in the white material of corn and of potatoes. The carbohydrates are formed of three elements--carbon, oxygen, and hydrogen. The use of these carbohydrates is to furnish to animal bodies either heat or energy or to enable them to store fat.

In the next place, let us look at the fat in plant food. This consists of the oil stored up in the seeds and other parts of the plant. The grains contain most of the oil. Fat is used by the animal to make heat and energy or to be stored away in the body.

The next animal food in the plant that we are to think about is the mineral matter. The ashes of a burnt plant furnish a common example of this mineral matter. The animal uses this material of the plant to make bone, teeth, and tissue.

The last thing that the plant furnishes the animal is water--just common water. Young plants contain comparatively large quantities of water. This is one reason why they are soft, juicy, and palatable. But, since animals get their water chiefly in another way, the water in feed stuffs is not important.

WHAT THESE COMPOUNDS DO IN THE BODY

Protein

1. Forms flesh, bone, blood, internal organs, hair, and milk. 2. May be used to make fat. 3. May be used for heat. 4. May be used to produce energy.

Carbohydrates

1. Furnish body heat. 2. Furnish energy. 3. Make fat.

Fat

1. Furnishes body heat. 2. Furnishes energy. 3. Furnishes body fat.

Mineral Matter

Furnishes mineral matter for the bones in the body.

Water

Supplies water in the body.

CHAPTER XI

FARM DAIRYING

SECTION LX. THE DAIRY COW

Success in dairy farming depends largely upon the proper feeding of stock. There are two questions that the dairy farmer should always ask himself: Am I feeding as cheaply as I can? and, Am I feeding the best rations for milk and butter production? Of course cows can be kept alive and in fairly good milk flow on many different kinds of food, but in feeding, as in everything else, there is an ideal to be sought.

What, then, is an ideal ration for a dairy cow? Before trying to answer this question the word ration needs to be explained. By ration is meant a sufficient quantity of food to support properly an animal for one day. If the animal is to have a proper ration, we must bear in mind what the animal needs in order to be best nourished. To get material for muscle, for blood, for milk, and for some other things, the animal needs, in the first place, food that contains protein. To keep warm and fat, the animal must, in the second place, have food containing carbohydrates and fats. These foods must be mixed in right proportions.

With these facts in mind we are prepared for an answer to the question, What is an ideal ration?

First, it is a ration that, without waste, furnishes both in weight and bulk of dry matter a sufficient amount of digestible, nutritious food.

Second, it is a ration that is comparatively cheap.

Third, it is a ration in which the milk-forming food (protein) is rightly proportioned to the heat-making and fat-making food (carbohydrates and fat). Any ration in which this proportion is neglected is badly balanced.

Now test one or two commonly used rations by these rules. Would a ration of cotton-seed meal and cotton-seed hulls be a model ration? No. Such a ration, since the seeds are grown at home, would be cheap enough. However, it is badly balanced, for it is too rich in protein; hence it is a wasteful ration. Would a ration of corn meal and corn stover be a desirable ration? This, too, since the corn is home-grown, would be cheap for the farmer; but, like the other, it is badly balanced, for it contains too much carbohydrate food and is therefore a wasteful ration.

A badly balanced ration does harm in two ways: first, the milk flow of the cow is lessened by such a ration; second, the cow does not profitably use the food that she eats.

The following table gives an excellent dairy ration for the farmer who has a silo. If he does not have a silo, some other food can be used in place of the ensilage. The table also shows what each food contains. As you grow older, it will pay you to study such tables most carefully.

FEED STUFFS	Dry matter	Protein	Carbohydrates	Fat
Cowpea hay = 15 pounds[1]	13.50	1.62	5.79	.16
Corn stover = 10 pounds	5.95	.17	3.24	.07
Corn ensilage = 30 pounds	6.27	.27	3.39	.21
Cotton-seed meal = 2 pounds	1.83	.74	.33	.24
Total = 57 pounds	27.55	2.80	12.75	.68

[Footnote 1: Alfalfa or clover hay may take the place of cowpea hay.]

=Care of the Cow.= As the cow is one of the best money-makers on the farm, she should, for this reason, if for no other, be comfortably housed, well fed and watered, and most kindly treated. In your thoughts for her well-being, bear the following directions in mind:

1. If you are not following a balanced ration, feed each day several different kinds of food. In this way you will be least likely to waste food.

2. Feed at regular hours. Cows, like people, thrive best when their lives are orderly.

3. Milk at regular hours.

4. Brush the udder carefully with a moist cloth before you begin to milk. Cleanliness in handling makes the milk keep longer.

5. Always milk in buckets or cups that have been scalded since the last using. The hot water kills the bacteria that collect in the dents or cracks of the utensil.

6. Never let the milk pail remain in the stable. Milk rapidly absorbs impurities. These spoil the flavor and cause the milk to sour.

7. Never scold or strike the cow. She is a nervous animal, and rough usage checks the milk flow.

SECTION LXI. MILK, CREAM, CHURNING, AND BUTTER

=Milk.= Milk is, as you know, nature's first food for mammals. This is because milk is a model food--it contains water to slake thirst, ash to make bone, protein to make flesh and muscle, and fat and sugar to keep the body warm and to furnish energy.

=The Different Kinds of Milk.= Whole, or unskimmed, milk, skimmed milk, and buttermilk are too familiar to need description. When a cow is just fresh, her milk is called colostrum. Colostrum is rich in the very food that the baby calf needs. After the calf is a few days old, colostrum changes to what is commonly known as milk.

A noticeable fact in this table is that skimmed milk differs from unskimmed mainly in the withdrawal of the fat. Hence, if calves are fed on skimmed milk, they should have in addition some food like corn meal to take the place of the fat withdrawn. A calf cannot thrive on skimmed milk alone. The amount of nourishing fat that a calf gets out of enough milk to make a pound of butter can be bought, in the form of linseed or corn meal, for a very small amount, while the butter-fat costs, for table use, a much larger sum. Of course, then, it is not economical to allow calves to use unskimmed milk. Some people undervalue skimmed milk; with the addition of some fatty food it makes an excellent ration for calves, pigs, and fowls.

Along with its dry matter, its protein, its carbohydrates, and its fats, milk and its products possess another most important property. This property is hard to describe, for its elements and its powers are not yet fully understood. We do, however, know certainly this much: milk and the foods made from it have power to promote health and favor growth in a more marked degree than any other foods. It is generally agreed that this is due to the health-promoting and health-preserving substances which are called vitamines. Men of science are working with much care to try to add to our knowledge of these vitamines, which have so marvelous an influence on the health of all animals. Unless food, no matter how good otherwise, contains these vitamines, it does not nourish the body nor preserve bodily health as it should. A complete lack of vitamines in our food would cause death. Since, then, milk and its products-- butter, cheese, curds--are rich in vitamines, these health-giving and health-preserving foods should form a regular part of each person's diet.

=Cream.= Cream is simply a mixture of butter-fat and milk. The butter-fat floats in the milk in little globe-shaped bodies, or globules. Since these globules are lighter than milk, they rise to the surface. Skimming the milk is a mere gathering together of these butter-fat globules. As most of the butter-fat is contained in the cream, pains should be taken to get all the cream from the milk at skimming time.

After the cream has been collected, it must be allowed to "ripen" or to "sour" in order that it may be more easily churned. Churning is only a second step to collect in a compact shape the fat globules. It often happens that at churning-time the cream is too warm for successful separation of the globules. Whenever this is the case the cream must be cooled.

=The Churn.= Revolving churns without inside fixtures are best. Hence, in buying, select a barrel or a square box churn. This kind of churn "brings the butter" by the falling of the cream from side to side as the churn is revolved. Never fill the churn more than one-third or one-half full of cream. A small churn is always to be avoided.

=Churning.= The proper temperature for churning ranges from 58?to 62?Fahrenheit. Test the cream when it is put into the churn. If it be too cold, add warm water until the proper temperature is reached; if too warm, add cold water or ice until the temperature is brought down to 62? Do not churn too long, for this spoils butter. As soon as the granules of butter are somewhat smaller than grains of wheat, stop the churn. Then draw off the buttermilk and at a temperature as low as 50?wash the butter in the churn. This washing with cold water so hardens the granules that they do not mass too solidly and thus destroy the grain.

=Butter.= The butter so churned is now ready to be salted. Use good fine dairy salt. Coarse barrel salt is not fit for butter. The salt can be added while the butter is still in the churn or after it is put upon the butter-worker. Never work by hand. The object of working is to get the salt evenly distributed and to drive out some of the brine. It is usually best to work butter twice. The two workings bring about a more even mixture of the salt with the butter and drive off more water. But one cannot be too particular not to overwork butter. Delicate coloring, attractive stamping with the dairy owner's special stamp, and proper covering with paper cost little and of course add to the ready and profitable sale of butter.

DAIRY RULES

Stable and Cows

1. Whitewash the stable once or twice each year; use land plaster, muck, or loam daily in the manure-gutters.

2. On their way to pasture or milking-place, do not allow the cows to be driven at a faster gait than a comfortable walk.

3. Give abundance of pure water.

4. Do not change feed suddenly.

5. Keep salt always within reach of each cow.

Milking

1. Milk with dry hands.

2. Never allow the milk to touch the milker's hands.

3. Require the milker to be clean in person and dress.

4. Milk quietly, quickly, thoroughly. Never leave a drop of milk in the cow's udder.

5. Do not allow cats, dogs, or other animals around at milking-time.

Utensils

1. Use only tin or metal cans and pails.

2. See that all utensils are thoroughly clean and free from rust.

3. Require all cans and pails to be scalded immediately after they are used.

4. After milking, keep the utensils inverted in pure air, and sun them, if possible, until they are wanted for use.

5. Always sterilize the churn with steam or boiling water before and after churning. This prevents any odors or bad flavors from affecting the butter. All cans, pails, and bottles should also be sterilized daily.

SECTION LXII. HOW MILK SOURS

On another page you have been told how the yeast plant grows in cider and causes it to sour, and how bacteria sometimes cause disease in animals and plants. Now you must learn what these same living forms have to do with the souring of milk, and maybe you will not forget how you can prevent your milk from souring. In the first place, milk sours because bacteria from the air fall into the milk, begin to grow, and very shortly change the sugar of the milk to an acid. When this acid becomes abundant, the milk begins to curdle. As you know, the bacteria are in air, in water, and in barn dust; they stick on bits of hay and stick to the cow. They are most plentiful, however, in milk that has soured; hence, if we pour a little sour milk into a pail of fresh milk, the fresh milk will sour very quickly, because we have, so to speak, "seeded" or "planted" the fresh milk with the souring germs. No one, of course, ever does this purposely in the dairy, yet people sometimes do what amounts to the same thing--that is, put fresh milk into poorly cleaned pails or pans, the cracks and corners of which are cozy homes for millions of germs left from the last sour milk contained in the vessel. It follows, then, that all utensils used in the dairy should be thoroughly scalded so as to kill all germs present, and particular care should be taken to clean the cracks and crevices, for in them the germs lurk.

In addition to this thorough cleansing with hot water, we should be careful never to stir up the dust of the barn just before milking. Such dusty work as pitching hay or stover or arranging bedding should be done either after or long before milking-time, for more germs fall into the milk if the air be full of dust.

To further avoid germs the milker should wear clean overalls, should have clean hands, and, above all, should never wet his hands with milk. This last habit, in addition to being filthy, lessens the keeping power of the milk. The milker should also moisten the parts of the cow which are nearest him, so that dust from the cow's sides may not fall into the milker's pail. For greater cleanliness and safety many milkmen curry their cows.

The first few streams from each teat should be thrown away, because the teat at its mouth is filled with milk which, having been exposed to the air, is full of germs, and will do much toward souring the other milk in the pail. Barely a gill will be lost by throwing the first drawings away, and this of the poorest milk too. The increase in the keeping quality of the milk will much more than repay the small loss. If these precautions are taken, the milk will keep several hours or even several days longer than milk carelessly handled. By taking these steps to prevent germs from falling into the milk, a can of milk was once kept sweet for thirty-one days.

The work of the germ in the dairy is not, however, confined to souring the milk. Certain kinds of germs give to the different sorts of cheeses their marked flavors and to butter its flavor. If the right germ is present, cheese or butter gets a proper flavor. Sometimes undesirable germs gain entrance and give flavors that we do not like. Such germs produce cheese or butter diseases. "Bitter butter" is one of these diseases. To keep out all unpleasant meddlers, thoroughly cleanse and scald every utensil.

EXERCISE

What causes milk to sour? Why do unclean utensils affect the milk? How should milk be cared for to prevent its souring? Prepare two samples, one carefully, the other carelessly. Place them side by side. Which keeps longer? Why?

SECTION LXIII. THE BABCOCK MILK-TESTER

It is not sufficient for a farmer or a dairyman to know how much milk each of his cows yields. He should also know how rich the milk is in butter-fat. Wide-awake makers of butter and cheese now buy milk, not by the pound or by the gallon, but by the amount of butter-fat contained in each pound or gallon of milk. A gallon of milk containing four and a half per cent of fat will consequently be worth more than a gallon containing only three per cent of fat. So it may happen that a cow giving only two gallons of milk may pay a butter-maker more than a cow giving three gallons of milk. Of course it is easy to weigh or measure the quantity of milk given by a cow, and most milkers keep this record; but until recent years there was no way to find out the amount of fat in a cow's milk except by a slow and costly chemical test. Dairymen could only guess at the richness of milk.

In 1890 Dr. S. M. Babcock of the Wisconsin Experiment Station invented a wonderful little machine that quickly and cheaply measures the fat in milk. Few machines are more useful. So desirous was Dr. Babcock of helping the farmers that he would not add to the cost of his machine by taking out a patent on his invention. His only reward has been the fame won by the invention of the machine, which bears his name. This most useful tester is now made in various sizes so that every handler of milk may buy one suited to his needs and do his own testing at very little cost.

The operation of the machine is very simple. Suppose that the members of the class studying this book have been asked to take a Babcock machine and test the milk of a small herd of cows. They can readily do so by following these directions:

While the milk is still warm from the first cow to be tested, mix it thoroughly by pouring it at least four times from one vessel to another. A few ounces of this mixed milk is then taken for a sample, and carefully marked with the name of the cow. A number is also put on the sample, and both the cow's name and the number entered in a notebook. A small glass instrument, called a pipette, comes with each machine. Put one end of the pipette into the milk sample and the other end into the mouth. Suck milk into the pipette until the milk comes up to the mark on the side of the pipette. As soon as the mark is reached, withdraw the pipette from the mouth and quickly press the forefinger on the mouth end. The pressure of the finger will keep the milk from running out. Then put the lower end of the pipette into one of the small long-necked bottles of the machine, and, lifting the finger, allow the milk to flow gently into the bottle. Expel all the milk by blowing through the pipette.

The next step is to add a strong, biting acid known as sulphuric acid to the test-bottle into which you have just put the milk. A glass marked to show just how much acid to use also comes with the machine. Fill this glass measure to the mark. Then pour the acid carefully into the test-bottle. Be sure not to drop any of the acid on your hands or your clothes. As the acid is heavier than the milk, it will sink to the bottom of the bottle. With a gentle whirling motion, shake the bottle until the two fluids are thoroughly mixed. The mixture will turn a dark brown and become very warm.

Now fill the other bottles in the same way with samples drawn from different cows. Treat all the samples precisely as you did the first. Do not forget to put on each sample the name of the cow giving the milk and on each test-bottle a number corresponding to the name of the cow.

You are now ready to put the test-bottles in the sockets of the machine. Arrange the bottles in the sockets so that the whirling frame of the machine will be balanced. Fit the cover on the machine and turn the handle slowly. Gradually gain in speed until the machine is whirled rapidly. Continue the turning for about seven minutes at the speed stated in the book of directions.

After this first turning is finished, pour enough hot water into each test-bottle to cause the fat to rise to the neck of the bottle. Re-cover the machine and turn for one minute. Again add hot water to each bottle until all the fat rises into the neck of the bottle and again turn one minute.

There remains now only the reading of the record. On the neck of each bottle there are marks to measure the amount of fat. If the fat inside the tube reaches only from the lowest mark to the second mark, then there is only one per cent of fat in this cow's milk. This means that the owner of the cow gets only one pound of butter-fat from each hundred pounds of her milk. Such a cow would not be at all profitable to a butter-seller. If the fat in another test-bottle reaches from the lowest mark to the fourth mark, then you put in your record-book that this cow's milk contains four per cent of butter-fat. This record shows that the second cow's milk yields four pounds of fat to every hundred pounds of milk. This cow is three times more valuable to a butter-maker than the first cow. In the same way add one more per cent for each higher mark reached by the fat. Four and one-half per cent is a good record for a cow to make. Some cows yield as high as five or six per cent but they do not generally keep up this record all the year.

[Illustration: FIG. 274. BABCOCK TESTER AND HOW TO USE IT The tester, acid, acid measure, test-bottle, and thermometer at bottom; filling the pipette on right; adding the acid and measuring the fat at top]

The Babcock tester shows only the amount of pure butter-fat in the milk. It does not tell the exact amount of finished butter which is made from 100 pounds of milk. This is because butter contains a few other things in addition to pure butter-fat. Finished and salted butter weighs on an average about one sixth more than the fat shown by the tester. Hence to get the exact amount of butter in every 100 pounds of milk, you will have to add one sixth to the record shown by the tester. Suppose, for example, you took one sample from 600 pounds of milk and that your test showed 4 per cent of fat in every 100 pounds of milk. Then, as you had 600 pounds of milk, you would have 24 pounds of butter-fat. This fat, after it has been salted and after it has absorbed moisture as butter does, will gain one sixth in weight. As one sixth of 24 is 4, this new 4 pounds must be added to the weight of the butter-fat. Hence the 600 pounds of milk would produce about 28 pounds of butter.

EXERCISE

1. Find the number of pounds of butter in 1200 pounds of milk that tests 3 per cent of butter-fat.

2. A cow yields 4800 pounds of milk in a year. Her milk tests 4 per cent of butter-fat. Find the total amount of butter-fat she yields. Find also the total amount of butter.

3. The milk of two cows was tested: one yielded in a year 6000 pounds of

milk that tested 3 per cent of fat; the other yielded 5000 pounds that tested 4 per cent. Which cow yielded the more butter-fat? What was the money value of the butter produced by each if butter-fat is worth twenty-five cents a pound?

CHAPTER XII

MISCELLANEOUS

SECTION LXIV. GROWING FEED STUFFS ON THE FARM

Economy in raising live stock demands the production of all "roughness" or roughage materials on the farm. By roughness, or roughage, of course you understand that bulky food, like hay, grass, clover, stover, etc., is meant. It is possible to purchase all roughage materials and yet make a financial success of growing farm animals, but this certainly is not the surest way to succeed. Every farm should raise all its feed stuffs. In deciding what forage and grain crops to grow we should decide:

1. The crops best suited to our soil and climate. 2. The crops best suited to our line of business. 3. The crops that will give us the most protein. 4. The crops that produce the most. 5. The crops that will keep our soil in the best condition.

1. The crops best suited to our soil and climate. Farm crops, as every child of the farm knows, are not equally adapted to all soils and climates. Cotton cannot be produced where the climate is cool and the seasons short. Timothy and blue grass are most productive on cool, limestone soils. Cowpeas demand warm, dry soils. But in spite of climatic limitations, Nature has been generous in the wide variety of forage she has given us.

Our aim should be to make the best use of what we have, to improve by selection and care those kinds best adapted to our soil and climate, and to secure, by better methods of growing and curing, the greatest yields at the least possible cost.

2. The crops best suited to our line of business. A farmer necessarily becomes more or less of a specialist; he gathers those kinds of live stock about him which he likes best and which he finds the most profitable. He should, on his

farm, select for his main crops those that he can grow with the greatest pleasure and with the greatest profit.

The successful railroad manager determines by practical experience what distances his engines and crews ought to run in a day, what coal is most economical for his engines, what schedules best suit the needs of his road, what trains pay him best. These and a thousand and one other matters are settled by the special needs of his road.

Ought the man who wants to make his farm pay be less prudent and less far-sighted? Should not his past failures and his past triumphs decide his future? If he be a dairy farmer, ought he not by practical tests to settle for himself not only what crops are most at home on his land but also what crops in his circumstances yield him the largest returns in milk and butter? If swine-raising be his business, how long ought he to guess what crop on his land yields him the greatest amount of hog food? Should a colt be fed on one kind of forage when the land that produced that forage would produce twice as much equally good forage of another kind? All these questions the prudent farmer should answer promptly and in the light of wise experiments.

3. The crops that will give us the most protein. It is the farmer's business to grow all the grass and forage that his farm animals need. He ought never to be obliged to purchase a bale of forage. Moreover, he should grow mainly those crops that are rich in protein materials, for example, cowpeas, alfalfa, and clover. If such crops are produced on the farm, there will be little need of buying so much cotton-seed meal, corn, and bran for feeding purposes.

4. The crops that produce the most. We often call a crop a crop without considering how much it yields. This is a mistake. We ought to grow, when we have choice of two crops, the one that is the best and the most productive on the farm. Average corn, for instance, yields on an acre at least twice the quantity of feeding-material that timothy does.

5. The crops that will keep our soil in the best condition. A good farmer should always be thinking of how to improve his soil. He wants his land to support him and to maintain his children after he is dead.

Since cowpeas, clover, and alfalfa add atmospheric nitrogen to the soil and at the same time are the best feeding-materials, it follows that these crops should hold an important place in every system of crop-rotation. By proper rotating, by proper terracing, and by proper drainage, land may be made to retain its fertility for generations.

EXERCISE

1. Why are cowpeas, clover, and alfalfa so important to the farmer?

2. What is meant by the protein of a food?

3. Why is it better to feed the farm crops to animals on the farm rather than to sell these crops?

SECTION LXV. FARM TOOLS AND MACHINES

The drudgery of farm life is being lessened from year to year by the invention or improvement of farm tools and machines. Perhaps some of you know how tiresome was the old up-and-down churn dasher that has now generally given place to the "quick-coming" churns. The toothed, horse-drawn cultivator has nearly displaced "the man with the hoe," while the scythe, slow and back-breaking, is everywhere getting out of the way of the mowing-machine and the horserake. The old heavy, sweat-drawing grain-cradle is slinking into the backwoods, and in its place we have the horse-drawn or steam-drawn harvester that cuts and binds the grain, and even threshes and measures it at one operation. Instead of the plowman's wearily making one furrow at a time, the gang-plows of the plains cut many furrows at one time, and instead of walking the plowman rides. The shredder and husker turns the hitherto useless cornstalk into food, and at the same time husks, or shucks, the corn.

The farmer of the future must know three things well: first, what machines he can profitably use; second, how to manage these machines; third, how to care for these machines.

The machinery that makes farming so much more economical and that makes the farmer's life so much easier and more comfortable is too complicated to be put into the hands of bunglers who will soon destroy it, and it is too costly to be left in the fields or under trees to rust and rot.

If it is not convenient for every farmer to have a separate tool-house, he should at least set apart a room in his barn, or a shed for storing his tools and machines. As soon as a plow, harrow, cultivator--indeed any tool or machine--has finished its share of work for the season, it should receive whatever attention it needs to prevent rusting, and should be carefully housed.

Such care, which is neither costly nor burdensome, will add many years to the life of a machine.

SECTION LXVI. LIMING THE LAND

Occasionally, when a cook puts too much vinegar in a salad, the dish becomes so sour that it is unfit to eat. The vinegar which the cook uses belongs to a large group of compounds known as acids. The acids are common in nature. They have the power not only of making salads sour but also of making land sour. Frequently land becomes so sour from acids forming in it that it will not bear its usual crops. The acids must then be removed or the land will become useless.

The land may be soured in several ways. Whenever a large amount of vegetable matter decays in land, acids are formed, and at times sourness of the soil results. Often soils sour because they are not well drained or because, from lack of proper tillage, air cannot make its way into the soil. Sometimes all these causes may combine to produce sourness. Since most crops cannot thrive on very sour soil, the farmer must find some method of making his land sweet again.

So far as we now know, liming the land is the cheapest and surest way of overcoming the sourness. In addition to sweetening the soil by overcoming the acids, lime aids the land in other ways: it quickens the growth of helpful bacteria; it loosens stiff, heavy clay soils and thereby fits them for easier tillage; it indirectly sets free the potash and phosphoric acid so much needed by plants; and it increases the capillarity of soils.

However, too much must not be expected of lime. Often a farmer's yield is so increased after he has scattered lime over his fields that he thinks that lime alone will keep his land fertile. This belief explains the saying, "Lime enriches the father but beggars the son." The continued use of lime without other fertilization will indeed leave poor land for the son. Lime is just as necessary to plant growth as the potash and nitrogen and phosphoric acid about which we hear so much, but it cannot take the place of these plant foods. Its duty is to aid, not to displace them.

We can tell by the taste when salads are too sour; it is more difficult to find out whether land is sour. There are, however, some methods that will help to determine the sourness of the soil.

In the first place, if land is unusually sour, you can determine this fact by a simple test. Buy a pennyworth of blue litmus paper from a drug store. Mix some of the suspected soil with a little water and bury the litmus paper in the mixture. If the paper turns red the soil is sour.

In the second place, the leguminous crops are fond of lime. Clover and vetch remove so much lime from the soil that they are often called lime plants. If clover and vetch refuse to grow on land on which they formerly flourished, it is generally, though not always, a sign that the land needs lime.

In the third place, when water grasses and certain weeds spring up on land, that land is usually acid, and lime will be helpful. Moreover, fields adjoining land on which cranberries, raspberries, blackberries, or gallberries are growing wild, may always be suspected of more or less sourness.

Four forms of lime are used on land. These, each called by different names, are as follows:

First, quicklime, which is also called burnt lime, caustic lime, builders' lime, rock lime, and unslaked lime.

Second, air-slaked lime, which is also known as carbonate of lime, agricultural lime, marl, and limestone.

Third, water-slaked, or hydrated, lime.

Fourth, land plaster, or gypsum. This form of lime is known to the chemists as sulphate of lime. Do not forget that this last form is never to be used on sour lands. We shall therefore not consider it further.

Air-slaked lime is simply quicklime which has taken from the air a gas called carbon dioxide. This is the same gas that you breathe out from your lungs.

Water-slaked lime is quicklime to which water has been added. In other words, both of these are merely weakened forms of quicklime. One hundred pounds of quicklime is equal in richness to 132 pounds of water-slaked lime and to 178 pounds of air-slaked lime. These figures should be remembered by a farmer when he is buying lime. If he can buy a fair grade of quicklime delivered at his railway station for $5.00 a ton, he cannot afford to pay more than $3.75 a ton for water-slaked lime, nor more than $2.75 for air-slaked lime of equal grade. Quicklime should always be slaked before it is applied to the soil.

As a rule lime should be spread broadcast and then harrowed or disked thoroughly into the soil. This is best done after the ground has been plowed. For pastures or meadows air-slaked lime is used as a top-dressing. When air-slaked lime is used it may be spread broadcast in the spring; the other forms should be applied in the fall or in the early winter.

SECTION LXVII. BIRDS

What do birds do in the world? is an important question for us to think about. First, we must gain by observation and by personal acquaintance with the living birds a knowledge of their work and their way of doing it. In getting this knowledge, let us also consider what we can do for our birds to render their work as complete and effective as possible.

Think of what the birds are doing on every farm, in every garden, and about every home in the land. Think of the millions of beautiful wings, of the graceful and attractive figures, of the cunning nests, and of the singing throats! Do you think that the whole service of the birds is to be beautiful, to sing charmingly, and to rear their little ones? By no means is this their chief service to man. Aside from these services the greatest work of birds is to destroy insects. It is one of the wise provisions of nature that many of the most brilliantly winged and most enchanting songsters are our most practical friends.

Not all birds feed on insects and animals; but even those that eat but a small amount of insect food may still destroy insects that would have damaged fruit and crops much more than the birds themselves do.

As to their food, birds are divided into three general classes. First, those that live wholly or almost wholly on insects. These are called insectivorous birds. Chief among these are the warblers, cuckoos, swallows, martins, flycatchers, nighthawks, whippoorwills, swifts, and humming-birds. We cannot have too many of these birds. They should be encouraged and protected. They should be supplied with shelter and water.

Birds of the second class feed by preference on fruits, nuts, and grain. The bluebird, robin, wood thrush, mocking-bird, catbird, chickadee, cedar-bird, meadow lark, oriole, jay, crow, and woodpecker belong to this group. These birds never fail to perform a service for us by devouring many weed seeds.

The third class is known as the hard-billed birds. It includes those birds which live principally on seeds and grain--the canary, goldfinch, sparrow, and some others.

Birds that come early, like the bluebird, robin, and redwing, are of special service in destroying insects before the insects lay their eggs for the season.

The robins on the lawn search out the caterpillars and cutworms. The chipping sparrow and the wren in the shrubbery look out for all kinds of insects. They watch over the orchard and feed freely on the enemies of the apple and other fruit trees. The trunks of these trees are often attacked by borers, which gnaw holes in the bark and wood, and often cause the death of the trees. The woodpeckers hunt for these appetizing borers and by means of their barbed tongues bring them from their hiding-places. On the outside of the bark of the trunk and branches the bark lice work. These are devoured by the nuthatches, creepers, and chickadees.

During the winter the bark is the hiding-place for hibernating insects, which, like plant lice, feed in summer on the leaves. Throughout the winter a single chickadee will destroy great numbers of the eggs of the cankerworm moth and of the plant louse. The blackbirds, meadow larks, crows, quail, and sparrows are the great protectors of the meadow and field crops. These birds feed on the army worms and cutworms that do so much injury to the young shoots; they also destroy the chinch bug and the grasshopper, both of which feed on cultivated plants.

A count of all the different kinds of animals shows that insects make up nine tenths of them. Hence it is easy to see that if something did not check their increase they would soon almost overrun the earth. Our forests and orchards furnish homes and breeding-places for most of these insects. Suppose the injurious insects were allowed to multiply unchecked in the forests, their numbers would so increase that they would invade our fields and create as much terror among the farmers as they did in Pharaoh's Egypt. The birds are the only direct friends man has to destroy these harmful insects. What benefactors, then, these little feathered neighbors are!

It has been estimated that a bird will devour thirty insects daily. Even in a widely extended forest region a very few birds to the acre, if they kept up this rate, would daily destroy many bushels of insects that would play havoc with the neighboring orchards and fields.

Do not imagine, however, that to destroy insects is the only use of birds. The day is far more delightful when the birds sing, and when we see them flit in and out, giving us a glimpse now and then of their pretty coats and quaint ways. By giving them a home we can surround ourselves with many birds, sweet of song and brilliant of plumage.

If the birds felt that man were a friend and not a foe, they would often turn to him for protection. During times of severe storm, extreme drought, or scarcity of food, if the birds were sufficiently tamed to come to man as their friend, as they do in rare cases now, a little food and shelter might tide them over the hard time and their service afterwards would repay the outlay a thousandfold. If the boys in your families would build bird-houses about the house and barn and in shade trees, they might save yearly a great number of birds. In building these places of shelter and comfort, due care must be taken to keep them clear of English sparrows and out of the reach of cats and bird-dogs.

Whatever we do to attract the birds to make homes on the premises must be done at the right time and in the right way. Think out carefully what materials to provide for them. Bits of string, linen, cotton, yarn, tow and other waste material, all help to induce a pair to build in the garden.

It is an interesting study--the preparation of homes for the birds. Trees may be pruned to make inviting crotches. A tangled, overgrown corner in the garden will invite some birds to nest.

Wrens, bluebirds, chickadees, martins, and some other varieties are all glad to set up housekeeping in man-made houses. The proper size for a bird-room is easily remembered. Give each room six square inches of floor space and make it eight inches high. Old, weathered boards should be used; or, if paint is employed, a dull color to resemble an old tree-trunk will be most inviting. A single opening near the top should be made two inches in diameter for the larger birds; but if the house is to be headquarters for the wren, a one-inch opening is quite large enough, and the small door serves all the better to keep out English sparrows.

The barn attic should be turned over to the swallows. Small holes may be cut high up in the gables and left open during the time that the swallows remain with us. They will more than pay for shelter by the good work they do in ridding the barn of flies, gnats, and mosquitoes.

SECTION LXVIII. FARMING ON DRY LANDS

Almost in the center of the western half of our continent there is a vast area in which very little rain falls. This section includes nearly three hundred million acres of land. It stretches from Canada on the north into Texas on the south, and from the Missouri River (including the Dakotas and western Minnesota) on the east to the Rocky Mountains on the west. In this great area farming has to be done with little water. This sort of farming is therefore called "dry-farming."

The soil in this section is as a rule very fertile. Therefore the difference between farming in this dry belt and farming in most of the other sections of our country is a difference mainly due to a lack of moisture.

As water is so scarce in this region two things are of the utmost importance: first, to save all the rain as it falls; second, to save all the water after it has fallen. To save the falling rain it is necessary for the ground to be in such a condition that none of the much-needed rain may run off. Every drop should go into the soil. Hence the farmer should never allow his top soil to harden into a crust. Such a crust will keep the rain from sinking into the thirsty soil. Moreover the soil should be deeply plowed. The deeper the soil the more water it can hold. The land should also be kept as porous as possible, for water enters a porous soil freely. The addition of humus in the form of vegetable manures will keep the soil in the porous condition needed. Second, after the water has entered the soil it is important to hold it there so that it may supply the growing crops. If the land is allowed to remain untilled after a rain or during a hot spell, the water in it will evaporate too rapidly and thus the soil, like a well, will go dry too soon. To prevent this the top soil should be stirred frequently with a disk or smoothing harrow. This stirring will form a mulch of dry soil on the surface, and this will hold the water. Other forms of mulch have been suggested, but the soil mulch is the only practical one. It must be borne in mind that this surface cultivation must be regularly kept up if the moisture is to be retained.

Some experiments in wheat-growing have shown how readily water might be saved if plowing were done at the right time. Wheat sowed on land that was plowed as soon as the summer crops were taken off yielded a very much larger return than wheat sowed on land that remained untilled for some time after the summer crops were gathered. This difference in yield on lands of the same fertility was due to the fact that the early plowing enabled the land to take up a sufficient quantity of moisture.

In addition to a vigilant catching and saving of water, the farmer in these dry climates must give his land the same careful attention that lands in other regions need. The seed-bed should be most carefully prepared. It should be deep, porous, and excellent in tilth. During the growing season all crops should be frequently cultivated. The harrow, the cultivator, and the plow should be kept busy. The soil should be kept abundantly supplied with humus.

Some crops need a little different management in dry-farming. Corn, for example, does best when it is listed; that is, planted so that it will come up three or four inches beneath the surface. If planted in this way, it roots better, stands up better, and requires less work.

Just as breeders study what animals are best for their climates, so farmers in the dry belt should study what crops are best suited to their lands. Some crops, like the sorghums and Kafir corn, are peculiarly at home in scantily watered lands. Others do not thrive. Experience is the only sure guide to the proper selection.

To sum up, then, farmers can grow good crops in these lands only when four things are done: first, the land must be thoroughly tilled so that water can freely enter the soil; second, the land must be frequently cultivated so that the water will be kept in the soil; third, the crops must be properly rotated so as to use to best advantage the food and water supply; fourth, humus must be freely supplied so as to keep the soil in the best possible condition.

SECTION LXIX. IRRIGATION

Irrigation is the name given to the plan of supplying water in large quantities to growing crops. Since the dawn of history this practice has been more or less followed in Asia, in Africa, and in Europe. The Spanish settlers in the southwestern part of America were probably the first to introduce this custom into our country. In New Mexico there is an irrigating trench that has been in constant use for three hundred years.

The most common source of water for irrigating purposes is a river or a smaller stream. Artesian wells are used in some parts of the country. Windmills are sometimes used when only a small supply of water is needed. Engines, hydraulic rams, and water-wheels are also employed. The water-wheel is one of the oldest and one of the most useful methods of raising water from streams. There are thousands of these in use in the dry regions of the West. Small buckets are fastened to a large wheel, which is turned by the current of a stream. As the wheel turns, the buckets are filled, raised, and then emptied into a trough called a flume. The water flows through the flume into the irrigating ditches, which distribute it as it is needed in the fields. In some parts of California and other comparatively dry sections, wells are sunk in or near the beds of underground streams, and then the water is pumped into ditches which convey it to the fields to be irrigated.

Engines are often used for pumping water from streams and transferring it to ditches or canals. The canals distribute the water over the land or over the growing crops.

None of these methods, however, can be used for watering very large areas of land. Hence, as the value of farm lands increased other methods were sought. Shrewd men began to turn longing eyes on the wide stretches of barren land in the West. They knew that these waste lands, seemingly so unfertile, would become most fruitful as soon as water was turned on them. Could water enough be found? New plans to pen up floods of water were prepared, and immense sums were spent in carrying out these plans. Enormous dams of cemented stone were thrown across the gorges in the foothills of the mountains. Behind these solid dams the water from the rains and the melting snow of the mountains was backed for miles, and was at once ready to change barrenness into fruitfulness. The stored water is led by means of main canals and cross ditches wherever it is needed, and countless acres have been brought under cultivation.

Water is generally applied either by making furrows for its passage through the fields or by flooding the land. The latter plan is the cheaper, but it can be used only on level lands. Where the land is somewhat irregular a checking system, as it is called, is used to distribute the water. It is taken from check to check until the entire field has been irrigated.

The furrow method is usually employed for fruits and for farm and garden crops. In many places the grass and grain crops are now supplied with water by furrows instead of by flooding.

Irrigated lands should be carefully and thoroughly tilled. The water for irrigation is costly, and should be made to go as far as possible. Good tillage saves the water. Moreover, all cultivated crops like corn, potatoes, and orchard and truck crops ought to be cultivated frequently to save the moisture, to keep the soil in fit condition, and to aid the bacteria in the soil. It was a wise farmer who said, "One does not need to grow crops many years in order to learn that nothing can take the place of stirring the soil."

METHODS OF IRRIGATING CROPS

Tree fruits. Water is conducted through very narrow furrows from three to five feet apart, and allowed to sink about four feet deep, and to spread under the ground. Then the supply is cut off. The object is to wet the soil deeply, and then by tillage to hold the moisture in the soil.

Small fruits. The common practice is to run water on each side of the row until the rows are soaked.

Potatoes. A thorough soaking is given the land before planting-time, and then no more than is absolutely necessary until blossoming-time. After the blossoms appear keep the soil moist until the crop ripens.

Garden crops. Any method may be employed, but the vital point is to cultivate the ground as early as it can be worked after it has been irrigated.

Meadows and alfalfa. Flooding is the most common method in use. The first irrigation comes early in the spring before growth has advanced much, and the successive waterings after the harvesting of each crop.

SECTION LXX. LIFE IN THE COUNTRY

As ours is a country in which the people rule, every boy and every girl ought to be trained to take a wide-awake interest in public affairs. This training cannot begin too early in life. A wise old man once said, "In a republic you ought to begin to train a child for good citizenship on the day of its birth."

Happy would it be for our nation if all the young people who live in the country could begin their training in good citizenship by becoming workers for these four things:

First, attractive country homes.

Second, attractive country schoolhouses and school grounds.

Third, good country schools.

Fourth, good roads.

If the thousands on thousands of pupils in our schools would become active workers for these things and continue their work through life, then, in less than half a century, life in the country would be an unending delight.

One of the problems of our day is how to keep bright, thoughtful, sociable, ambitious boys and girls contented on the farm. Every step taken to make the country home more attractive, to make the school and its grounds more enjoyable, to make the way easy to the homes of neighbors, to school, to post-office, and to church, is a step taken toward keeping on the farm the very boys and girls who are most apt to succeed there.

Not every man who lives in the country can have a showy or costly home, but as long as grass and flowers and vines and trees grow, any man who wishes can have an attractive house. Not every woman who is to spend a lifetime at the head of a rural home can have a luxuriously furnished home, but any woman who is willing to take a little trouble can have a cozy, tastefully furnished home--a home fitted with the conveniences that diminish household drudgery. Even in this day of cheap literature, all parents cannot fill their children's home with papers, magazines, and books, but by means of school and Sunday-school libraries, by means of circulating book clubs, and by a little self-denial, earnest parents can feed hungry minds just as they feed hungry bodies.

Agricultural papers that arouse the interest and quicken the thought of farm boys by discussing the best, easiest, and cheapest ways of farming; journals full of dainty suggestions for household adornment and comfort; illustrated papers and magazines that amuse and cheer every member of the family; books that rest tired bodies and open and strengthen growing minds--all of these are so cheap that the money reserved from the sale of one hog will keep a family fairly supplied for a year.

If the parents, teachers, and pupils of a school join hands, an unsightly, ill-furnished, ill-lighted, and ill-ventilated school-house can at small cost be changed into one of comfort and beauty. In many places pupils have persuaded their parents to form clubs to beautify the school grounds. Each father sends a man or a man with a plow once or twice a year to work a day on the grounds. Stumps are removed, trees trimmed, drains put in, grass sowed, flowers, shrubbery, vines, and trees planted, and the grounds tastefully laid off. Thus at scarcely noticeable money cost a rough and unsightly school ground gives place to a charming school yard. Cannot the pupils in every school in which this book is studied get their parents to form such a club, and make their school ground a silent teacher of neatness and beauty?

Life in the country will never be as attractive as it ought to be until all the roads are improved. Winter-washed roads, penning young people in their own homes for many months each year and destroying so many of the innocent pleasures of youth, build towns and cities out of the wreck of country homes. Can young people who love their country and their country homes engage in a nobler crusade than a crusade for improved highways?

APPENDIX

SPRAYING MIXTURES

FOR BITING INSECTS

DRY PARIS GREEN

Paris green 1 lb. Lime or flour 4 to 16 lb.

WET PARIS GREEN

Paris green 1/4 to 2 lb. Lime 1/4 to 1/2 lb. Water 50 gal.

FOR SOFT-BODIED SUCKING INSECTS

KEROSENE EMULSION

Hard soap (in fine shavings) 1/2 lb. Soft water 1 gal. Kerosene 2 gal.

Dissolve soap in boiling water, add kerosene to the hot water, churn with spraying pump for at least ten minutes, until the mixture changes to a creamy, then to a soft, butterlike, mass. This gives three gallons of 66-per-cent oil emulsion, which may be diluted to the strength desired. To get 15-per-cent oil emulsion add ten and one-half gallons of water.

FOR FUNGOUS DISEASES

COPPER SULPHATE

Copper sulphate 1 lb. Water 18 to 25 gal.

Use only before foliage opens, to kill wintering spores.

BORDEAUX MIXTURE

Copper sulphate (bluestone) 4 to 5 lb. Lime (good, unslaked) 5 to 6 lb. Water 50 gal.

Dissolve the copper sulphate (bluestone) in twenty-five gallons of water. Slake the lime slowly so as to get a smooth, thick cream. Never cover the lime with too much water. After thorough slaking add twenty-five gallons of water. When the lime and the bluestone have dissolved, pour the two liquids into a third vessel. Be sure that each stream mixes with the other before either enters the vessel. Strain through a coarse cloth.

Mix fresh for each time. Use for molds and fungi generally. Apply in fine spray with a good nozzle.

BORDEAUX-PARIS-GREEN MIXTURE

Ordinary Bordeaux mixture 50 gal. Paris green 4 oz. to 2 lb.

Use for both fungi and insects on apple, potato, etc.

BORDEAUX-ARSENATE-OF-LEAD MIXTURE

Ordinary Bordeaux mixture 50 gal. Arsenate of lead 2 to 3 lb.

Used for fungous and insect enemies of the potato, and of the apple when bitter rot is troublesome.

COMMERCIAL LIME-SULPHUR ARSENATE OF LEAD

Commercial lime-sulphur 1-1/2 gal. Arsenate of lead 2 to 3 lb. Water 50 gal.

Use for spraying apples.

AMMONIACAL COPPER CARBONATE

Copper carbonate 5 oz. Ammonia (26?Baum? about 3 pt. Water 50 gal.

Dissolve the copper carbonate in the smallest possible amount of ammonia. This solution may be kept in stock and diluted to the proper strength as needed.

Use this instead of the Bordeaux mixture after the fruit has reached half or two thirds of the mature size. It leaves no spots as does the lime-sulphur wash or the Bordeaux mixture.

SPRAYS FOR BOTH FUNGOUS AND INSECT PESTS

HOME-MADE LIME-SULPHUR WASH

Lime 20 lb. Sulphur 15 lb. Water 50 gal.

The lime, the sulphur, and about half of the water required are boiled together for forty-five minutes in a kettle over a fire, or in a barrel or other suitable tank by steam, strained, and then diluted to 50 gallons. This is the wash regularly used against the San Jose scale. It may be substituted for Bordeaux mixture when spraying trees in the dormant state. Commercial lime-sulphur may also be used in place of this homemade wash. Use one gallon of the commercial lime-sulphur to nine gallons of water in the dormant season.

SELF-BOILED LIME-SULPHUR WASH

The self-boiled lime-sulphur wash is a combination of lime and sulphur boiled only by the heat of the slaking lime, and is used chiefly for summer spraying on peaches, plums, cherries, etc. as a substitute for the Bordeaux mixture.

Lime 8 lb. Sulphur 6 to 8 lb. Water 50 gal.

The lime should be placed in a barrel and enough water poured on it to start it slaking and to keep the sulphur off the bottom of the barrel. The sulphur, which should first be worked through a sieve to break up the lumps, may then be added, and, finally, enough water to slake the lime into a paste. Considerable stirring is necessary to prevent caking on the bottom. After the violent boiling which accompanies the slaking of the lime is over, the mixture should be diluted ready for use, or at least enough cold water added to stop the cooking. From five to fifteen minutes are required for the process. If the hot mass is permitted to stand undiluted as a thick paste, a liquid is produced that is injurious to peach foliage and, in some cases, to apple foliage.

The mixture should be strained through a sieve of twenty meshes to the inch in order to remove the coarse particles of lime, but all the sulphur should be worked through the strainer.

GLOSSARY

To enable young readers to understand the technical words necessarily used in the text only popular definitions are given.

=Abdomen=: the part of an insect lying behind the thorax.

=Acid=: a chemical name given to many sour substances. Vinegar and lemon juice owe their sour taste to the acid in them.

=Adult=: a person, animal, or plant grown to full size and strength.

=Ammonia= (ammonium): a compound of nitrogen readily usable as a plant food. It is one of the products of decay.

=Annual=: a plant that bears seed during the first year of its existence and then dies.

=Anther=: the part of a stamen that bears the pollen.

=Atmospheric nitrogen=: nitrogen in the air. Great quantities of this valuable plant food are in the air; but, strange to say, most plants cannot use it directly from the air, but must take it in other forms, as nitrates, etc. The legumes are an exception, as they can use atmospheric nitrogen.

=Available plant food=: food in such condition that plants can use it.

=Bacteria=: a name applied to a number of kinds of very small living beings, some beneficial, some harmful, some disease-producing. They average about one twenty-thousandth of an inch in length.

=Balanced ration=: a ration made up of the proper amounts of carbohydrates, fats, and protein, as explained in text. Such a ration avoids all waste of food.

=Biennial=: a plant that produces seed during the second year of its existence and then dies.

=Blight=: a diseased condition in plants in which the whole or a part of a plant withers or dries up.

=Bluestone=: a chemical; copper sulphate. It is used to kill fungi, etc.

=Bordeaux Mixture=: a mixture invented in Bordeaux, France, to destroy disease-producing fungi.

=Bud= (noun): an undeveloped branch.

=Bud= (verb): to insert a bud from the scion upon the stock to insure better fruit.

=Bud variation=: occasionally one bud on a plant will produce a branch differing in some ways from the rest of the branches; this is bud variation. The shoot that is produced by bud variation is called a sport.

=Calyx=: the outermost row of leaves in a flower.

=Cambium=: the growing layer lying between the wood and the bark.

=Canon=: the shank bone above the fetlock in the fore and hind legs of a horse.

=Carbohydrates=: carbohydrates are foods free from nitrogen. They make up the largest part of all vegetables. Examples are sugar, starch, and cellulose.

=Carbolic acid=: a chemical often used to kill or prevent the growth of germs, bacteria, fungi, etc.

=Carbon=: a chemical element. Charcoal is nearly pure carbon.

=Carbon disulphide=: a chemical used to kill insects.

=Carbonic acid gas=: a gas consisting of carbon and oxygen. It is produced by breathing, and whenever carbon is burned. It is the source of the carbon in plants.

=Cereal=: the name given to grasses that are raised for the food contained in their seeds, such as corn, wheat, rice.

=Cobalt=: a poisonous chemical used to kill insects.

=Cocoon=: the case made by an insect to contain its larva or pupa.

=Commercial fertilizer=: an enriching plant food bought to improve soil.

=Compact=: a soil is said to be compact when the particles are closely packed.

=Concentrated=: when applied to food the word means that it contains much feeding value in small bulk.

=Contagious=: a disease is said to be contagious when it can be spread or carried from one individual to another.

=Cross=: the result of breeding two varieties of plant together.

=Cross pollination=: the pollination of a flower by pollen brought from a flower on some other plant.

=Croup=: the top of the hips.

=Culture=: the art of preparing ground for seed and raising crops by tillage.

=Curb disease=: a swelling on the back part of the hind leg of a horse just behind the lowest part of the hock joint. It generally causes lameness.

=Curculio=: a kind of beetle or weevil.

=Dendrolene=: a patented substance used for catching cankerworms.

=Digestion=: the act by which food is prepared by the juices of the body to be used by the blood.

=Dormant=: a word used to describe sleeping or resting bodies,--bodies not in a state of activity.

=Drainage=: the process by which an excess of water is removed from the land by ditches, terraces, or tiles.

=Element=: a substance that cannot be divided into simpler substances.

=Ensilage=: green foods preserved in a silo.

=Evaporate=: to pass off in vapor, as a fluid often does; to change from a solid or liquid state into vapor, usually by heat.

=Exhaustion=: the state in which strength, power, and force have been lost. When applied to land, the word means that land has lost its power to produce well.

=Fermentation=: a chemical change produced by bacteria, yeast, etc. A common example of fermentation is the change of cider into vinegar.

=Fertility=: the state of being fruitful. Land is said to be fertile when it produces well.

=Fertilization=: the act which follows pollination and enables a flower to produce seed.

=Fetlock=: the long-haired cushion on the back side of a horse's leg just above the hoof.

=Fiber=: any fine, slender thread or threadlike substance, as the rootlets of plants or the lint of cotton.

=Filter=: to purify a liquid, as water, by causing it to pass through some substance, as paper, cloth, screens, etc.

=Formalin=: a forty per cent solution of a chemical known as formaldehyde. Formalin is used to kill fungi, bacteria, etc.

=Formula=: a recipe for the making of a compound; for example, fertilizer or spraying compounds.

=Fungicide=: a substance used to kill or prevent the growth of fungi; for example, Bordeaux Mixture or copper sulphate.

=Fungous=: belonging to or caused by fungi.

=Fungus= (plural =fungi=): a low kind of plant life lacking in green color. Molds and toadstools are examples.

=Germ=: that from which anything springs. The term is often applied to any very small organism or living thing, particularly if it causes great effects such as disease, fermentation, etc.

=Germinate=: to sprout. A seed germinates when it begins to grow.

=Girdle=: to make a cut or groove around a limb or tree.

=Glacier=: an immense field or stream of ice formed in the region of constant snow and moving slowly down a slope or valley.

=Globule=: a small particle of matter shaped like a globe.

=Glucose=: a kind of sugar very common in plants. The sugar from grapes, honey, etc. is glucose. That from the sugar cane is not.

=Gluten=: a vegetable form of protein found in cereals.

=Graft=: to place a living branch or stem on another living stem so that it may grow there. It insures the growth of the desired kind of plant.

=Granule=: a little grain.

=Gypsum=: land plaster.

"=Head back=": to cut or prune a tree so as to form its head, that is, the place where the main trunk first gives off its branches.

=Heredity=: the resemblance of offspring to parent.

=Hibernating=: to pass the winter in a torpid or inactive state in close quarters.

=Hock=: the joint in the hind leg of quadrupeds between the leg and the shank. It corresponds to the ankle in man.

=Host=: the plant upon which a fungus or insect is preying.

=Humus=: the portion of the soil caused by the decay of animal or vegetable matter.

=Hybrid=: the result of breeding two different kinds of plants together.

=Hydrogen=: a chemical element. It is present in water and in all living things.

=Individual=: a single person, plant, animal, or thing of any kind.

=Inoculate=: to give a disease by inserting the germ that causes it in a healthy being.

=Insectivorous=: anything that eats insects.

=Kainit=: salts of potash used in making fertilizers.

=Kernel=: a single seed or grain, as a kernel of corn.

=Kerosene emulsion=: see Appendix.

=Larva= (plural =larv?=): the young or immature form of an insect.

=Larval=: belonging to larva.

=Layer=: to propagate plants by a method similar to cutting, but differing from cutting in that the young plant takes root before it is separated from the parent plant.

=Legume=: a plant belonging to the family of the pea, clover, and bean; that is, having a flower of similar structure.

=Lichen=: a kind of flowerless plant that grows on stones, trees, boards, etc.

=Loam=: an earthy mixture of clay and sand with organic matter.

=Magnesia=: an earthy white substance somewhat similar to lime.

=Magnify=: to make a thing larger in fact or in appearance; to enlarge the appearance of a thing so that the parts may be seen more easily.

=Membrane=: a thin layer or fold of animal or vegetable matter.

=Mildew=: a cobwebby growth of fungi on diseased or decaying things.

=Mold=: see mildew.

=Mulch=: a covering of straw, leaves, or like substances over the roots of plants to protect them from heat, drought, etc., and to preserve moisture.

=Nectar=: a sweetish substance in blossoms of flowers from which bees make honey.

=Nitrate=: a readily usable form of nitrogen. The most common nitrate is saltpeter.

=Nitrogen=: a chemical element, one of the most important and most expensive plant foods. It exists in fertilizers, in ammonia, in nitrates, and in organic matter.

=Nodule=: a little knot or bump.

=Nutrient=: any substance which nourishes or promotes growth.

=Organic matter=: substances made through the growth of plants or animals.

=Ovary=: the particular part of the pistil that bears the immature seed.

=Ovipositor=: the organ with which an insect deposits its eggs.

=Oxygen=: a gas present in the air and necessary to breathing.

=Particle=: any very small part of a body.

=Perennial=: living through several years. All trees are perennial.

=Petal=: a single leaf of the corolla.

=Phosphoric acid=: an important plant food occurring in bones and rock phosphates.

=Pistil=: the part of the blossom that contains the immature seeds.

=Pollen=: the powdery substance borne by the stamen of the flower. It is necessary to seed production.

=Pollination=: the act of carrying pollen from stamens to pistils. It is usually done by the wind or by insects.

=Porosity=: the state of having small openings or passages between the particles of matter.

=Potash=: an important part of plant foods. The chief source of potash is kainit, muriate of potash, sulphate of potash, wood ashes, and cotton-hull ashes.

=Propagate=: to cause plants or animals to increase in number.

=Protein=: the name of a group of substances containing nitrogen. It is one of the most important of feeding stuffs.

=Pruning=: trimming or cutting parts that are not needed or that are injurious.

=Pulverize=: to reduce to a dustlike state.

=Pupa=: an insect in the stage of its life that comes just before the adult condition.

=Purity= (of seed): seeds are pure when they contain only one kind of seed and no foreign matter.

=Ration=: a fixed daily allowance of food for an animal.

=Raupenleim=: a patented sticky substance used to catch the cankerworm.

=Resistant=: a plant is resistant to disease when it can ward off attacks of the disease; for example, some varieties of the grape are resistant to the phylloxera.

=Rotation= (of crops): a well-arranged succession of different crops on the same land.

=Scion=: a shoot, sprout, or branch taken to graft or bud upon another plant.

=Seed bed=: the layer of earth in which seeds are sown.

=Seed selection=: the careful selection of seed from particular plants with the object of keeping or increasing some desirable quality.

=Seedling=: a young plant just from the seed.

=Sepal=: one of the leaves in the calyx.

=Set=: a young plant for propagation.

=Silo=: a house or pit for packing away green food for winter use so as to exclude air and moisture.

=Sire=: father.

=Smut=: a disease of plants, particularly of cereals, which causes the plant or some part of it to become a powdery mass.

=Spike=: a lengthened flower cluster with stalkless flowers.

=Spiracle=: an air opening in the body of an insect.

=Spore=: a small body formed by a fungus to reproduce the fungus. It serves the same use as seeds do for flowering plants.

=Spray=: to apply a liquid in the form of a very fine mist by the aid of a spraying pump for the purpose of killing fungi or insects.

=Stamen=: the part of the flower that bears the pollen.

=Stamina=: endurance.

=Sterilize=: to destroy all the germs or spores in or on anything. Sterilizing is often done by heat or chemicals.

=Stigma=: the part of the pistil that receives the pollen.

=Stock=: the stem or main part of a tree or plant. In grafting or budding the scion is inserted upon the stock.

=Stover=: as used in this book the word means the dry stalks of corn from which the ears have been removed.

=Subsoil=: the soil under the topsoil.

=Sulphur=: a yellowish chemical element; brimstone.

=Taproot=: the main root of a plant, which runs directly down into the earth to a considerable depth without dividing.

=Terrace=: a ridge of earth run on a level around a slope or hillside to keep the land from washing.

=Thorax=: the middle part of the body of an insect. The thorax lies between the abdomen and the head.

=Thermometer=: an instrument for measuring heat.

=Tillage=: the act of preparing land for seed, and keeping the ground in a proper state for the growth of crops.

=Transplant=: a plant grown in a bed with a view to being removed to other soil; a technical term used by gardeners.

=Tubercle=: a small, wart-like growth on the roots of legumes.

=Udder=: the milk vessel of a cow.

=Utensil=: a vessel used for household purposes.

=Variety=: a particular kind. For example, the Winesap, Bonum, 苊 op, etc., are different varieties of apples.

=Ventilate=: to open to the free passage of air.

=Virgin soil=: a soil which has never been cultivated.

=Vitality= (of seed): vitality is the ability to grow. Seed are of good vitality if a large per cent of them will sprout.

=Weathering=: the action of moisture, air, frost, etc. upon rocks.

=Weed=: a plant out of place. A wheat plant in a rose bed or a rose in the wheat field would be regarded as a weed, as would any plant growing in a place in which it is not wanted.

=Wilt= (of cotton): a disease of cotton in which the whole plant droops or wilts.

=Withers=: the ridge between the shoulder bones of a horse, at the base of the neck.

=Yeast=: a preparation containing the yeast plant used to make bread rise, etc.